D1871437

THE EVOLUTION OF PLANTS

THE EVOLUTION OF PLANTS

By

DUKINFIELD HENRY SCOTT

GYAN

Published by

GYAN

Gyan Publishing House
5, Ansari Road
Daryaganj, New Delhi-110002
Phone: 011-47034999, 9811692060
E-mail: books@gyanbooks.com

Distribution Network
gyanbooks.com
India, USA, Canada, UK, Australia, France

© Publisher

All rights reserved. No part of this work may be reproduced, stored, adapted, or transmitted in any form or by any means, electronic, mechanical, photocopying, micro-filming, recording or otherwise, or translated in any language, without the prior written permission of the copyright owner and the publisher.

ISBN : 978-81-212-6438-9 (PB)
First Published, 1911

2nd Impression 2022

Printed at: Gyan Press, Delhi.

The book is sold subject to the condition that it shall not, by way of trade or otherwise, be lent, resold, hired out, or otherwise circulated without the prior publisher's written consent.

THE EVOLUTION OF PLANTS
Author: DUKINFIELD HENRY SCOTT

THE EVOLUTION OF PLANTS

BY

DUKINFIELD HENRY SCOTT

1911

CONTENTS

THE EVOLUTION OF PLANTS

CHAPTER I

INTRODUCTION — THE DARWINIAN THEORY — NATURE OF THE EVIDENCE — THE FOSSIL RECORD

THE doctrine of Evolution is an attempt to answer the question, "What was the origin of all the multitude of forms of living animals and plants which now people the earth?" This question, so natural to ask, was long regarded by the majority of competent naturalists as insoluble.

Though we cannot, in the nature of things, have any direct knowledge of the origin of nature species (*i. e.* .the different kinds of wild plants and animals), yet there are a great many familiar and often very distinct forms of living things, of which we do know the history, in whole or in part. These are our cultivated plants whole or in part. These are our cultivated plants and domestic animals. We know, for example (to confine ourselves to the plants), that our Roses and Tulips, Potatoes and Parsnips, Wheat and Oats, with their numerous varieties and races, all came originally from wild species, often extremely different from the garden or field plants which man has created out of them. We will not go into the question *how* the cultivated forms came from the wild types —that is a long and involved story, always difficult and often impossible to unravel in detail. For example, the origin of Wheat, the most important to us of all cultivated plants, has long been a mystery, and is not yet fully cleared up. The Wheats have now been traced, as Dr. Stapf informs us, to four primitive types. The most important of the wild species (the original of the Hard Wheats and other forms) has only been recognised within the last few years as a weed growing on the slopes of Mount Hermon in Palestine; the wild form of the Soft Wheats is still unknown.

But, whatever the difficulties, we at any rate know that all cultivated plants have come from pre-existing wild forms by descent with modification. This, however, is Evolution, and we arrive at the conclusion that all those plants of which we know the origin have been evolved from other, different plants which existed before them.

Cultivated forms are not of the same nature as wild species, for the conditions under which they have arisen are very different, but there is sufficient analogy to justify the inference that the wild forms also, which arose without man's interference and of which he has never had the chance of observing the development, had a similar origin, by descent with modification, from older types, these again from still older, and so on, further and further back into the past. This line of argument is due to Darwin, who first taught the world to believe in Evolution, though the idea is perhaps as old as human thought. Darwin used these words: "At the commencement of my observations it seemed to me probable that a careful study of domesticated animals and cultivated plants would offer the best chance of making out this obscure problem. Nor have I been disappointed. . . . I have invariably found that our knowledge, imperfect though it be, of variation under domestication, afforded the best and safest clue" (*Origin of Species*, p. 3).

Important chapters in the *Origin of Species* are devoted to this subject, which was worked out fully in Darwin's later book on *Animals and Plants under Domestication.*

The Darwinian theory is not only a statement of Evolution, it involves an explanation of Evolution, and without this explanation Darwin's work would not have carried conviction, any more than that of his predecessors, nor have exercised the vast influence which it has done on the thought of the world.

When the theory now known as Darwinism was first made public, on July 1, 1858, the title of the joint papers by Darwin and Wallace, then read before the Linnean Society, was: "On the tendency of Species to form Varieties, and on the Perpetuation of Varieties and Species by Natural Means of Selection." When Darwin's "Abstract" of his theory appeared in the following year it bore the title: "On the Origin of Species by means of Natural Selection or the Preservation of favoured Races in the Struggle for Life." The essential feature of the Darwinian theory is the selection, by natural means, of such variations as are advantageous to the plant or animal concerned. To quote Darwin's words: "As many more individuals of each species are born than can possibly survive; and as, consequently, there is a frequently recurring struggle for existence, it follows that any being if it vary however slightly in any manner profitable to itself, under the complex and sometimes varying conditions of life, will have a better chance of surviving, and thus be *naturally selected*. From the strong principle of inheritance, any selected variety will tend to propagate its new and modified form" (*Origin of Species*, p. 3).

"This preservation of favourable individual differences and variations, and the destruction of those which are injurious, I have called Natural Selection, or the Survival of the Fittest" (*ibid.*, p. 63).

Darwin liked the term "Selection" because it is constantly used in all works on breeding. He quotes an agricultural writer, Youatt, who spoke of the principle of Selection as "that which enables the agriculturalist not only to modify the character of his flock, but to change it altogether" (*ibid.*, p. 23). The same holds good for the improvement of garden plants and field crops; the principle in all cases has been to select for breeding the *best* animals or plants of each kind, *i. e.* those which best show the particular characters desired by the breeder or gardener, such as fine wool in sheep, size and flavour in fruit, or beauty of colour in flowers.

"Man," said Darwin, "selects only for his own good: Nature only for that of the being which she tends" (*ibid.*, p. 65). "If it profit a plant to have its seeds more and more widely disseminated by the wind, I can see no greater difficulty in this being effected through natural selection, than in the cotton-planter increasing and improving by selection the down in the pods on his cotton-trees" (*ibid.*, p. 67). Natural

Selection is, of course, a metaphorical expression; in Nature it is no conscious being that selects, but it is all the circumstances of life, such as soil, climate and associated organisms, which determine the survival of those varieties which are best adapted to them.

Variations, under which Darwin expressly included the differences which distinguish one individual from another, occur in every direction, harmful, indifferent or beneficial; it is on the continual selection for survival of the favourable variations that Evolution depends. Thus, for the first time, the Darwinian theory gave a natural explanation of the process of Evolution, and, at the same time, of the great mystery of adaptation. The universal characteristic of all living things is that their whole structure is a system of mechanisms and devices by which they are enabled to maintain their own life and leave descendants behind them. The whole of Biology bears witness to adaptation; as striking cases among plants, familiar to all, we may mention the beautiful mechanism for securing cross-fertilisation by insects, in the flowers of an Orchid or a Sage; the means of dispersal possessed by seeds or seed-vessels in the Balsam, the Bedstraw or the Dandelion; the climbing organs in the Virginian Creeper, the Clematis or the Ivy.

Natural Selection appears to be the only theory at present in the field, which can be said to give at all a satisfactory explanation, by means of natural causes, of the origin of adaptations, which once formed the basis of Paley's famous argument from Design. Sir William Thiselton-Dyer has said on this point: "The thought of each age is the foundation of that which follows. Darwin was an admirer of Paley, a member of his own college. He swept in the whole of Paley's teleology, simply dispensing with its supernatural explanation."

Teleology is the doctrine that everything in Nature serves some end; applied to living things it is identical with the idea of adaptation. We must remember, however, that the end served, according to Darwinian doctrine, is always the benefit of the creature itself, whether animal or plant, not the benefit of man. The two things, it is true, often work together, for man himself determines, to a considerable extent, what shall survive and what shall perish. Wheat, for example, from a few scattered localities in the East, has overspread a great part of the world, to the exclusion of most of the wild Flora, simply because it has become the food of man. To be of use to man is now about the best asset in the struggle for existence to countless plants and animals.

The following words give Darwin's own estimate of the importance of Selection.

"Slow though the process of selection may be, if feeble man can do much by artificial selection, I can see no limit to the amount of change, to the beauty and complexity of the coadaptations between all organic beings, one with another and with their physical conditions of life, which may have been affected in the long course of time through nature's power of selection, that is by the survival of the fittest" (*Origin of Species*, p. 85).

The theory of Natural Selection has been much attacked, both on its first publication and in recent years; many of the opponents of the theory, however, have failed to grasp its meaning. "The great engine of Natural Selection," as Mr. Francis Darwin said in 1908, "is taunted now-a-days, as it was fifty years ago, with being merely a negative power" — that is, people fancy that it can do no more than get rid of the unfit. If, however, the comparison with artificial selection holds good, it is evident that the power is a positive one, capable of creating new forms, with organs perfectly adapted to their own needs, just as the cultivated forms created through selection by man are perfectly adapted to his requirements.

Darwin himself never claimed for Natural

Selection that it was the sole agent in Evolution; he regarded it as "the most important, but not the exclusive means of modification" (*Origin of Species*, p. 4). It is probable that other causes have played a considerable part.

I do not, however, propose to pursue the subject of the causes of Evolution. My purpose has only been to show how Darwin, by discovering in Natural Selection a true cause, though not the sole cause of Evolution, at once made that doctrine credible, and ultimately secured its universal acceptance.

The object of this book is to try to trace historically the course which the evolution of plants has actually followed — a most difficult problem, but one on which the work of the last few years has thrown light at several points. We shall confine ourselves to those groups of plants for which the evidence is the most satisfactory; these groups are, generally speaking, the highest and also the most familiar.

The evidence on which we have to rely comes partly from the comparative study of plants now living. A vast amount of work has been done, during the last two centuries, in finding out the relationships of plants and thus building up a natural system of classification. In so far as the attempt is successful, the result is to group plants into families, of which the members, if

Evolution is true, are related to each other in the same way as the members of a human family, though more distantly. Plants belonging to the same natural family will at any rate be more closely akin to each other than to members of other families; the families belonging to the same natural class will be more nearly related than those of different classes, and so on.

There are, however, great difficulties; we cannot always be sure that our groups are natural, and even if they are, we are still a long way from being able to trace their descent. The comparative method, applied to the study of the structure and life-histories of plants, has accomplished a great deal. The great Hofmeister, sixty years ago, established, once for all, the main lines of relationship between the Flowering Plants and the Higher Cryptogams (Ferns and their allies), and to a certain extent, between the latter and the Mosses; his work still forms the foundation of all our knowledge of the affinities of these great groups of plants. But, in questions of descent, all work based on the comparison of living forms leaves us in uncertainty, because the plants we are comparing are all, so far as we know, of equal antiquity; all belong as it were, to one generation. The comparative work of Hofmeister gave us very good reason for believing that Conifers and their allies were descended from the higher Cryptogams; it is, however, obvious that they were not descended from any Cryptogams now living, nor, in all probability, from any plants at all closely resembling them; Ferns and Club-mosses have changed in the course of ages as well as Conifers, though perhaps not to the same extent. The utmost we can get out of the comparison of living plants is the conclusion that the ancestors of one group (which we usually call the higher) were more or less like the members of another group (which we usually call the lower).

The idea of "higher" and "lower" is, however, very likely to lead us wrong. Many people suppose that Evolution means progress from the lower to the higher, from the simple to the complex, and are therefore apt to assume that a simple group of organisms is likely to represent the ancestry of a more complex, related group; very often, however, the reverse is the truth.

For example, a beginner might very naturally suppose that the Duckweeds, Flowering Plants as simple as Liverworts, with no regular distinction between stem and leaf, represent a primitive, ancestral stage in the evolution of the higher families of Flowering Plants; all botanists, however, are agreed that the Duckweeds are really degenerate water-forms, degraded from higher plants in consequence of their mode of life, and not primitive or ancestral at all. Here, I believe, there is no serious difference of opinion, but there are a great many cases in which botanists are divided on the question whether a simple form represents an old and primitive type or a mere reduction from something higher.

One of the most important questions of the kind is that of the relation of the Higher Cryptogams (Ferns and their allies) to the Bryophytes (Mosses and their allies). The Fern group is admittedly "higher" than the Moss group, and most botanists have believed that the former was derived, if not from Bryophytes, at least from plants of that type. This view is now much shaken, and it is beginning to appear more probable that the Higher Cryptogams are a more ancient and primitive group than the Bryophytes, which would seem to owe their origin to reduction from some higher type (see Chapter VIII).

Even if we take the very simplest organisms, the Bacteria, minute cells in which not even a nucleus can be demonstrated with certainty, it is quite doubtful whether they represent early, primitive forms, or degradations of higher organisms — the latter view, indeed, appears the more probable.

There is, however, as we saw at the beginning, one class of cases where we can hardly go wrong;

we are fairly safe in believing that cultivated plants came from the related wild forms, and not the other way about. Even this is not always quite certain; in following up the history of the Wheats, the possibility has often had to be considered that an apparently natural form might really be an old cultivated form run wild. We can only be certain of tracing a cultivated plant to its origin if we know its history. This brings us to our main point, the necessity for historical testimony; the analogy of cultivated plants shows us that we can only hope to gain any certain knowledge of the course of descent of plants if we know something of their past history; we need the most direct evidence we can get.

To a considerable extent such evidence is available. Preserved in the records of the rocks, invaluable though fragmentary chapters from the evolutionary history have come down to us. The importance and interest of the fossil record is recognised by every one as regards the Animal Kingdom. The monstrous Reptiles, the "Dragons of the Air" and the toothed birds, from the Secondary Rocks, the great Sloths and Mammoths of later days, are subjects of popular knowledge; many people are aware of the light that has been thrown on Evolution by the fossil history of the horse and the elephant.

Fossil plants are less familiar, but, in their

importance as historical documents, they are not unworthy to be compared with the animal record. If in some respects the botanist is less fortunate than his zoological colleague, in others he has the advantage. Though there is nothing in plants quite like the skeleton, internal or external, of animals, so well adapted for preservation and so valuable as evidence, yet in knowledge of outward form and anatomical structure the fossil-botanist has the best of it. Besides the impressions and casts of the stems, leaves, etc., which are the best known kinds of plant-fossils, we have in certain formations the still more valuable petrified specimens, in which the mineral substance, originally in solution, has so thoroughly permeated the tissues as to preserve their minute structure. We are thus able, for example, to study by means of thin sections the microscopic anatomy of many plants of the Coal Measures, with almost the same accuracy as if our specimens had just been gathered in the garden, instead of having lain buried in the earth for some millions of years.

We cannot always expect to get evidence as good as this, and every one knows that the fossil record is imperfect; the surprising thing is that it tells us so much. The subject of the Evolution of Plants will be treated in this book with constant reference to the fossil evidence.

As it is impossible to cover the whole ground, only the higher groups of plants are considered; for this part of the Vegetable Kingdom the fossil record is always available and always important.

The questions which will be considered are:

The Evolution of the true Flowering Plants or Angiosperms (Chapter II and III).

The Evolution of the Seed-plants generally (Chapter IV).

The Evolution of the great groups of the higher Cryptogams, *i. e.* of those Spore-plants which share with the Seed-plants the possession of a vascular system (wood and bast) (Chapters V to VII).

A table of geological strata is given for reference, in which the relative thickness of the various formations (down to the Cambrian) as they occur in Britain is roughly indicated on a scale of about 1 inch to 20,000 feet. A few words of explanation may be added, as regards those formations to which we shall have to refer in the following pages.

The Tertiary formations represent, geologically speaking, the modern period of the record. So far as plants are concerned, no really great change occurred during Tertiary times — that is to say, the same classes of plants which now constitute the Flora of the world flourished throughout the period, and about the same proportions as at present. The Tertiary period is very interesting to botanists because it covers

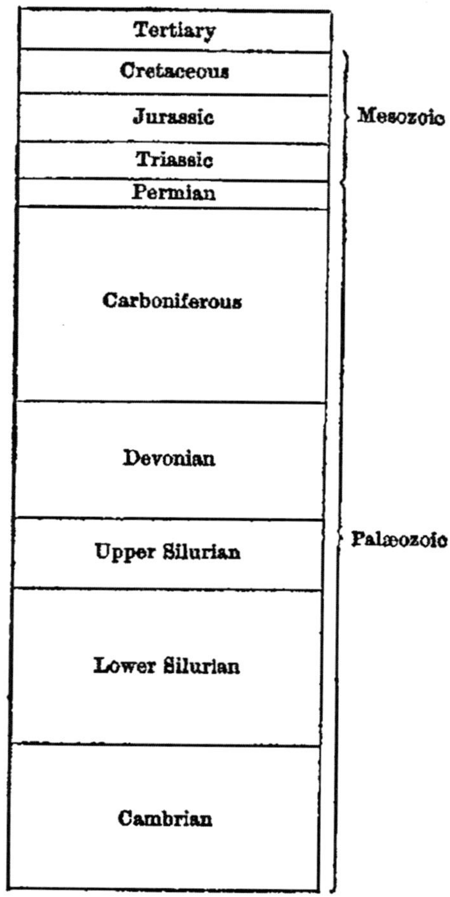

the evolution of many of the families of the higher Flowering Plants; at present, however, the time is not ripe for dealing with these questions. In the South of England the Tertiaries are very well developed—the London basin, for example, is occupied by strata of this age, the Bagshot Sands, the London Clay, and the Woolwich and Reading beds. In the Isle of Wight and on the Hampshire coast, Tertiary deposits of later date are well shown.

In the Cretaceous formation we reach the great Secondary or Mesozoic period. The geological boundaries, however, do not correspond here to any important change in the vegetation. Many of the modern types extend through the Upper, and some down to the Lower Cretaceous. In the latter, however, we find a more antique Flora, with Gymnosperms (Conifers and their allies) as the dominant class; the Chalk, so characteristic of South-eastern England, may serve to represent the Upper Cretaceous, while the Lower Greensand, well developed in Kent, Surrey and Sussex, is a good example of the Lower Cretaceous strata. At the base of the Cretaceous is an important formation—the Wealden—so named from the Weald of the three counties just mentioned. It is a fresh-water formation, particularly rich in fossil plants of the typical Mesozoic Flora (see below, p. 56).

The same vegetation, often abundantly represented, is found throughout the Jurassic formation, which includes the Purbeck Beds, the various Oolitic strata which form so large a part of the Midlands, and the Lias, famous for its fossil Reptiles. The Lower Oolites of the Yorkshire coast are extremely rich in plant-remains and have been called the "Oolitic Coal Measures."

To the Trias belongs the greater part of the "New Red Sandstone," which covers large tracts of country in the Midland and Western counties, reaching the South Devon coast, where it forms the red cliffs of Sidmouth. The plant-fossils are not numerous in England, but more frequent on the Continent. They are a little different in character from the Jurassic Flora, and some show an approach to the Palæozoic vegetation.

The Palæozoic period begins, as we descend, with the Permian, a formation represented, for example, by the Magnesian Limestone of Northern England; the red cliffs west of the Exe in Devonshire are also said to belong to it. In many parts of the Continent and especially in Russia, whence it takes its name the Permian is a great formation, and its lower beds rich in fossil plants. The transition from the Mesozoic Flora is practically lost; in the Permian we find ourselves at once among typical Palæozoic Vegetation, of the same general character as that of the Coal Measures—gigantic Club-mosses and Horsetails, with Gymnosperms and primitive Seed-plants of classes now unknown, characterise this famous Permo-carboniferous Flora, the most important of all in the history of the Vegetable World.

The Carboniferous consists in England of three main divisions: The Coal Measures, the Millstone Grit, and the Carboniferous Limestone.

The Coal Measures, consisting largely of the buried remains of vast swampy forests, are the richest of all formations in fossil plants, in all conditions of preservation. A peculiar feature of certain English coal-beds (belonging to the Lower Coal Measures of Lancashire and Yorkshire) is the presence in the coal itself of calcareous concretions called coal-balls, which are packed full of petrified fragments of plants, preserved with astonishing perfection. Coal-balls also occur in the roof of the seams, and contain a somewhat different Flora from that of the seam itself. Similar coal-balls are known to occur in the coal-fields of Germany and Austria.

In France, the Upper Coal Measures and Permian are especially rich in petrified specimens, usually preserved in silica.

Throughout the whole series, impressions and casts of plants are abundant and of great value in showing their external characters.

The Upper Carboniferous includes the whole of the Coal Measures (Upper, Middle and Lower), together with the underlying Millstone Grit. The Lower Carboniferous is chiefly represented in England by the Mountain or Carboniferous Limestone of the Lennine chain and the Mendip Hills, and in Southern and Central Scotland by the Calciferous Sandstones. There are considerable differences between the Flora of the Upper and that of the Lower Carboniferous, the latter approaching the Devonian in character. Plants of much interest and of quite peculiar types are found in the Calciferous Sandstones of Scotland.

The Devonian, so well represented in the country from which it derives its name, as well as by the old Red Sandstone of Scotland and elsewhere, is of extreme interest botanically, for it contains the oldest Land Flora of which we at present have any satisfactory knowledge. In our own islands it is only at certain places in Ireland that the Devonian is at all rich in fossil plants, but in North America a considerable Flora of Devonian age is known.

The plants do not on the whole differ very widely from those of the Lower Carboniferous, though some are of course peculiar. They are already very highly organised plants; a vast history must lie behind them, of which, as yet, we know practically nothing. At present it is only the Flora of the *Upper* Devonian which is at all well known.

The vast masses of ancient rock known as the Silurian and the Cambrian do not come into consideration in this book, for we have but little knowledge of land-plants from these strata; supposed discoveries of Silurian Ferns and Club-mosses are still open to doubt, though it appears certain that vascular plants must have existed at least as early. Some well-preserved seaweeds occur in Silurian and even, perhaps, in Cambian rocks, but the subject of the history of the Algæ, though very interesting, does not fall within the limits we have laid down.

Broadly speaking, there are only four really well-characterised Floras of the past, so far as we know—

I. The Tertiary and Upper Cretaceous; the modern Flora, or Era of the Angiosperms.

II. The Mesozoic Flora, or Era of the Gymnosperms.

III. The later Palæozoic Flora, or Era of the primitive Seed-plants and Cryptogams.

IV. The early Palæozoic Flora, or Era of the Algæ.

The last-mentioned is too little known to be really comparable to the others.

We are, here, only interested in the first three of these Floras. Our object will be, so far as the fragmentary evidence permits, to trace back the main lines of descent of the higher classes of plants, to Palæozoic times.

CHAPTER II

THE EVOLUTION OF THE FLOWERING PLANTS—THE PROBLEM

TEN years ago, Professor Vines, in his address to the Botanical Section of the British Association, was at the pains to take a census of the Vegetable Kingdom, according to the best statistics available at the time. A census of that kind is of course very different from the census we are expecting this year; it is a count of species, not of individuals. Botanists differ very much as to the limits of species; for example, in Hooker's *Students' Flora* only four British species of *Rubus* (the Blackberry and Raspberry genus) are recognised, while in Druce's *List of British Plants* 133 species are given! However, that is happily an extreme case, and as the object of a plant-census is to get at proportional rather than absolute numbers, the different views taken of species do not so very much matter. Neither does the fact that many new species have been described since the year 1900 affect the question.

Making a free use of Professor Vines's statistics, and rounding off his figures, we get the following results:

Angiosperms or Flowering Plants . .	103,000
Gymnosperms (*e. g.* Conifers)	2,500
Vascular Cryptogams (*e. g.* Ferns) . .	3,500
Mosses and Liverworts	7,500
Fungi and Bacteria	40,000
Lichens	5,500
Algæ (*e. g.* Seaweeds)	14,000
	176,000

Thus, of all known plants more than four-sevenths belong to the Angiosperms or Flowering Plants. The figures bring home to us in a statistical form what every one knows—that the Angiosperms or Flowering Plants are beyond all question the dominant class of the Vegetable Kingdom as it now exists. I speak of the Angiosperms as the Flowering Plants although that term is often used to include the Gymnosperms (Conifers and their allies) also. We do not, however, naturally speak of a fir-cone as a "flower," and in this case popular custom has seized on an essential distinction, for, so far as Gymnosperms now living are concerned, their cones have not much in common with true flowers, and the two classes are wide as the poles asunder, though still confused in many old-fashioned botany books.

The dominance of the Angiosperms in the Flora of the modern world is shown not only by the mere number of species, but still more by the importance of the part they play. As forest trees some Conifers can compete with them; occasionally Bracken-fern or Bog-moss may prevail over considerable areas; otherwise it is the Angiosperms which cover the face of the earth with vegetation. They have adapted themselves to every climate, from the High Alps, where *Ranunculus glacialis* grows at a height of over 14,000 feet in the Bernese Oberland, surrounded by perpetual snow, to the hot deserts of Central Africa or America, where gigantic Euphorbias and Cacti flourish in a rainless wilderness.

The Angiosperms, besides covering the surface of the earth with forests, prairies and meadows, have fitted themselves to every niche where plant-growth is possible.

In the tropics and in damp forests elsewhere, hundreds of species grow as *epiphytes* on trees and shrubs. Most of the showy Orchids in hothouses are of this epiphytic nature, and so are most members of the Pineapple family. Sometimes epiphytes adapt themselves in extraordinary ways to their peculiar mode of life. In some Orchids, the leaves are not developed, and the roots have taken their place, turning green and flattening themselves out over the bark of the tree on which they grow like the frond of a Liverwort. On the other hand, the Old Man's Beard of Western South America (*Tillandsia usneoides*) has lost its roots altogether, and hangs loose on the branches, in long grey tufts, like a lichen. An East Indian epiphyte (*Dischidia Rafflesiana*) has converted its leaves into pitchers, which act as "natural flower-pots," collecting soil brought by ants and the water that drips from the tree. The epiphyte sends down its roots into the pitchers, to feed on the soil which has accumulated in them.

The "natural flower-pots" of *Dischidia* must not be confounded with the pitchers of the better-known pitcher-plants such as *Nepenthes;* here the pitchers serve as traps for insects, for these plants, like our Sundew and the Venus's Fly-trap, are carnivorous, reversing the usual order of nature, and feeding on animals instead of animals feeding on them.

This is a curious example of the enterprise shown by Angiosperms; we have another in the parasitic plants of this class, which emulate Fungi, and live at the expense of some other plant on which they prey. Some parasites are only half-and-half robbers, making some of their food for themselves, and taking the rest from their victim. This is the case with green parasites, such as the Mistletoe, the best known of all; also the Eyebright and Money (*Rhinanthus cristagalli*), which are parasitic on roots and may do a good deal of harm; there is a saying that "The farmer who has money in his fields has none in his pocket." But others, like the Broom-rape and the Dodder, have lost their green colour and live altogether on the unhappy plants which they infest. The Dodder has practically no leaves or roots; it looks like a mere tangle of twine, on the furze-bush or other victim, until it begins to flower, and its true nature is revealed. This degradation of the vegetative body of the parasite in consequence of its living on the fruits of other plants' labours, goes much further in some tropical forms. In the famous *Rafflesia Arnoldi*, a parasite on Sumatran vines, the flower is the biggest in the world, a yard in diameter, but nothing else is to be seen. Stem, leaves and roots have all disappeared and are represented only by a web of threads, like the spawn of a fungus, burrowing in the substance of the plant attacked. Thus on occasion, a member of the highest class of the Vegetable Kingdom can descend to the level of a fungus in order to gain some advantage in the struggle for life. The flowers, however, still betray the real nature of the plant.

Other Flowering Plants resemble in their mode of life those Fungi that grow on the decaying remains of other plants. An example is the Bird's Nest Orchis of our woods (*Neottia nidus-avis*). Here, however, the business is more complicated, for the plant grows in close association with a real fungus, which helps it to get its food. This taking of a fungus into partnership is a very common practice, both among Orchids and other plants, but usually the Angiospermous partner (unlike the Bird's Nest Orchis) retains its green leaves, and with them a greater degree of independence.

These few examples of peculiar ways of living among Flowering Plants have been mentioned to show the wonderful adaptability displayed by the highest class of the Vegetable Kingdom—this has no doubt been one of the conditions of its success. The cases of insect-eating and parasitism are particularly interesting, because, so far as I know, no members of any other class of plants, above the rank of Fungi, have adopted these particular means of gaining a livelihood. At least this seems to be true at the present day, though we do not know what may have gone on in this way among the plants of the past.

Before leaving the subject of the adaptability of Angiosperms we must say a word about water-plants. As we all know, a great many Flowering Plants live in fresh water, some with floating leaves, like the Water-lily, some floating altogether, like the Duckweed, and some submerged altogether, like the Hornwort (*Ceratophyllum*). Such plants hold their own very well and play a more important part in fresh-water lakes, ponds and rivers than the Algæ themselves. The water Flowering Plants do not, however, always have it all their own way, for I am told that in some lakes in the Hebrides, the old-world Horsetails are now turning out the more advanced aquatics such as Reeds and Rushes.

The water-plants, especially those that live quite submerged, have their structure a good deal simplified compared with land-plants, because their requirements are so different. The simplification often extends to the flowers, and some botanists have imagined that certain water-plants are really primitive members of the Angiosperms. As a rule this is very unlikely, for in many cases we can trace quite clearly the families of land-plants to which they belong. Most, at any rate, of the aquatic Angiosperms have taken to the water from the land, and altered their structure accordingly. An extreme case is found in the Duckweeds referred to in the first chapter, plants which have lost almost all distinction between leaf and stem, and are reduced to a tiny frond which in *Wolffia* is even without roots. The flowers, too, are extremely simple. These minute water-weeds, which have lost all resemblance to ordinary Flowering Plants, are extremely successful and have spread over the fresh waters of the whole world. By simplifying their structure they have become able to compete on equal terms with the fresh-water Algæ, and often to beat them in their own domain; they have "stooped to conquer." Another family of water Angiosperms, which has descended to the level of lower plants, is that of the Podostemaceæ, which flourish on rocks in the rushing torrents and waterfalls of tropical streams. They too have become completely transformed, and have come to resemble anything but Flowering Plants—sometimes Mosses or Liverworts, sometimes Lichens and sometimes Algæ. In the case of this family, there is thus a great variety of forms, all alike strange and unfamiliar among Angiosperms.

In the sea alone the Angiosperms have met with but little success. On our own coasts we have our one marine Flowering Plant, the

Grasswrack (*Zostera*); in tropical seas there are a few more, but on the whole the flower is a device which has not proved to be well fitted for marine life, to which, indeed, it is wholly alien in its origin and history. In the sea things still go on in the old way, as they did before flowers were invented. Byron's words in his lines to the Ocean: "Such as Creation's dawn beheld, thou rollest now," are true, in a measure, of its Flora also.

It is, however, interesting to note that the Angiosperms alone, of all the higher groups of plants, appear to have had any effect at all on the Flora of the sea. We have no marine Moss, no marine Fern, no marine Gymnosperm; it is only when we come to the highest class of all, the Angiosperms, and probably to rather advanced families among them, that we find some few invaders of the Ocean.

So far as the land is concerned, at all events, the Angiosperms have most effectually entered into their kingdom; they are almost everywhere dominant, and under very nearly all conditions of life have put in their claim and taken possession. They are capable of infinite variety, ranging in dimensions from a frond a millimetre across (the Duckweed *Wolffia*) to trees 100 metres in height (*Eucalyptus*), and in complexity from a mere group of simple cells to the elaborate combination of tissue-systems in the stem of a tropical climber.

Yet, geologically speaking, the Angiosperms are a modern class of plants. At the time when our coal-forests spread so widely over the world they were, for all we know, not in existence; in fact, not a trace of them is found for many ages later; it is high up in the Secondary rocks that the first Angiosperms appear. Thus, so far as we can judge, their whole course of evolution lay within a period fully open to investigation by the palæontologist, and belonged indeed to quite the later part of the fossil record. The question, what was the origin of this great modern class of plants, which has prevailed over all competitors, is evidently of fascinating interest, and should not be of hopeless difficulty; yet the problem has always been the great puzzle of the botanical evolutionist, and it is only well within the present, still youthful century that any light upon it has appeared.

Darwin, writing in 1879, to Sir Joseph Hooker, said: "The rapid development as far as we can judge of all the higher plants within recent geological times is an abominable mystery," and so it has remained until a year or two ago. Even now the clue that has been found still seems to some uncertain, but it is at least well worth our following.

The subject of the geological history of the Angiosperms themselves is a difficult one, because the evidence, though abundant, is generally dubious, most of the fossils being impressions of leaves, from which it is often impossible to determine with any certainty the family to which the plant belonged. The whole question needs investigating afresh, but in the meantime a few general conclusions may be given.

Throughout the whole of the Tertiary period Angiosperms were abundant; in fact, they were dominant all through those long ages, much as they are now. They were a varied class of plants throughout, and a great number of our living families have been recognised with more or less certainty; Monocotyledons and Dicotyledons are alike represented, the latter, as at present, being much the more numerous.

To a certain extent the same was the case in the latter part of the Secondary period. The Upper Cretaceous rocks, of an age more or less corresponding to that of the Chalk, have yielded a good many remains of Angiosperms, which at that time were already the leading class of plants, forming, in most places, the great bulk of the Flora. The Cretaceous Angiosperms were on the whole quite ordinary members of their class, so far as the evidence enables us to judge. Birches, Beeches, Oaks, Walnuts, Planes, Maples, Hollies and Ivies, with many other familiar trees and shrubs, have been described from Upper Cretaceous rocks; even if we cannot always trust the identification, the facts are at any rate sufficient to prove the presence of a great variety of Dicotyledons, in habit quite similar to many of those now living. A certain number of Monocotyledons are also found; in the case of the great family of Palms, the evidence appears to be quite conclusive, for leaves, fruits, and stems with their structure preserved are all well known from Cretaceous rocks.

The oldest recorded fossil of the Palm kind appears to be a Cocoa-nut (or similar fruit) from the Cretaceous rocks of France, of an age nearly corresponding to our own Upper Greensand. Thus the great family of the Palms is certainly one of the oldest known among the Angiosperms, though there were plenty of Dicotyledons of the same age.

In the Upper Cretaceous beds of Japan, a number of petrified plant-remains, with their structure preserved, have lately been discovered by Dr. Marie Stopes and Prof. Fujii. In this Flora Angiosperms and Gymnosperms were almost equally numerous, to judge from the few species already described. Among the former is a fossil flower probably belonging to the Lily family.

When we get down to the Lower Cretaceous (of which our Lower Greensand is an example) we find ourselves at the beginning of the Angiospermous history. Their remains are very rare in most beds of this age, but at two places, one in Portugal and the other in the State of Virginia, they are found in considerable numbers. They resemble various living plants, such as Poplars, Willows, Fig-trees, and Laurels; Monocotyledons are also present. In the beds just below, Angiosperms are altogether absent, so we seem to have got back here to their first appearance, and, indeed, there are no older trustworthy records of this class of plants in any part of the world.

This apparently sudden appearance of quite well-developed Flowering Plants is still, perhaps, the greatest difficulty in the record of evolution. The Angiosperms must have had a previous history of some kind, and must, we may assume, have been derived from some older group of plants. Can we form any idea what group this was? This is the question which the work of the last few years has done something towards answering, and we will go on in the next chapter to explain what has been discovered about it.

In order to make the subject as plain as possible, it will be necessary first to go a little more fully into the characters which distinguish the Angiosperms, so that we may be better able to compare them with other classes of plants.

The first great characteristic of Angiosperms is that they alone, among plants now living, possess true flowers. What is a flower? In ordinary life our idea of a flower is associated with bright colour and often with a sweet scent. But we now know that colour and scent are not there merely to give pleasure to us, but that they are of important service to the plant, by attracting the visits of insects. Sometimes, in fact, neither the colour nor the scent is by any means pleasing to ourselves, for they may serve to attract such insects as carrion flies, which have very different tastes from our own; this is the case in the *Rafflesia*, mentioned above, and many of the Arum family. The visits of insects (and in a few cases of other animals, such as Humming-Birds) are chiefly of use to the plant, as Darwin showed, by bringing about cross-fertilisation between different flowers or different individuals. In other words, a typical flower is an organ of sexual reproduction, adapted to crossing by means of animal visitors, especially insects. The numerous cases in which the "short circuit" method of self-fertilisation has been resorted to, do not affect the main conclusion, for in many of these cases the flower has evidently once been adapted for crossing, though now self-fertilised. In the Bee-Orchis, for example, this is obvious; the flower has all the beautiful and elaborate arrangements which in other related Orchids ensure crossing, but the stalks of the pollen masses are long and weak, so that they dangle down and are blown by the wind against the stigma of the same flower.

Popular language, in its use of the word "flower," has grasped what is really essential to that type of flower which is prevalent in the existing Flora and has been prevalent for long ages before, as the geological record shows. It is probable that the close relation to insect-life has been the chief condition determining the evolution of Angiosperms and giving them their supremacy among living vegetation.

The higher families of insects, which are chiefly concerned in the fertilisation of flowers, appeared during the Secondary period. The Bees and Wasps (Hymenoptera) do not appear, so far as we know, before the Upper Oolite; the earliest of the Lepidoptera (Butterflies and Moths) are of about the same age. Thus both these groups, according to our present knowledge, appeared only a little before the true Flowering Plants. The insects of the Carboniferous age were chiefly Cockroaches and Dragon-flies (see fig. 12, p. 105). The latter reached an enormous size, one species having a body 14 inches long, with a spread of wings of quite 2 feet.

When the Angiosperms came in so suddenly, as it seems to us, in Cretaceous times, the whole face of the world was changed, and flowers like those with which we are now familiar everywhere began to appear. This, the greatest change which the Kingdom of Plants has ever known—almost comparable to the advent of Man in the Animal record—doubtless chiefly depended on the simultaneous development of the higher forms of insect life. This holds good, though, as we all know, there are many Flowering Plants which are not fertilised by insects but by the wind. It is quite likely that the chief wind-fertilised families, such as the catkin-bearing trees, were derived from insect-fertilised ancestors, but this is a matter of theory at present, and we will not pursue the question.

The arrangement of the different organs in Angiospermous flowers is remarkably constant, in spite of the infinite variety of form which flowers show. Where all the parts are present, we have on the outside the floral envelopes or perianth, the leaves of which may either be all more or less alike as in a Lily or a Magnolia, or, as is more often the case, may be in two distinct sets, an outer, the calyx, commonly green and serving for a protection to the bud, and an inner, the corolla, commonly composed of brightly-coloured leaves, the petals. This is the case in most of our familiar flowers, especially those of the plants which have two seed-leaves (Dicotyledons). Either the whole perianth, or the corolla, as the case may be, is the conspicuous part, in cases where the flower is attractive to insects by its colour.

Within the perianth come the stamens, ranged in a single circle as in the Primrose, or in a double circle as in the Geranium, or numerous and spirally disposed as in the Buttercup. Lastly, in the middle of the whole flower, stands the pistil, consisting of one or more carpels, either free or united in various ways, and containing the ovules or young seeds.

This typical structure is common to Dicotyledons and Monocotyledons; it is subject to every degree of variation, in the number of the organs and in their freedom from, or union to, one another. Thus in the Buttercup all parts of the flower are free and separate; in the Strawberry and many other plants of the Rose family, all parts are free, except the calyx, in which the sepals are partly joined together. In a Primrose the petals of the corolla are also united to form a tube, and the stamens are joined to the corolla. The stamens in most flowers are separate from one another, but in the Mallow they are all joined by their stalks in a single ring, while in the St. John's Wort they are united in groups. The carpels are more often united than free, anong themselves; in the Orchids, stamen and pistil are fused to form the column. The pistil is always in the middle of the flower, but in many families that part of it, the ovary, which contains the ovules, is sunk below the level of the calyx; the ovary is then called inferior, as in the Hemlock family, the Campanulas and many more.

It is not my purpose to go any further into the modifications of the flower; my object is to point out that the possession of a flower, in the sense explained, is characteristic of Angiosperms, and that we may naturally expect to find at least the beginnings of a flower in any group of plants from which the Angiosperms are likely to have been derived.

Before leaving the subject it must be recalled to mind that the flower in many Angiosperms is much simpler than in the typical cases dealt with above. Often there is no corolla (Knotgrass; Nettle); often there is no perianth at all (Spurge; Willow). What is more important, the flower may be uni-sexual, only the stamens or only the pistil being present, as is the case in all our catkin-bearing trees, and many of the Arum family. Where, however, the individual flowers are very simple, we almost always find a number of them combined together, as in an Arum or a Willow, the whole group or inflorescence answering more or less the purpose of a single flower. The question whether the simpler flowers are primitive, or reduced from more perfect forms, has been much discussed; certainly reduction is in many cases the true explanation of simplicity, and it may be that this holds good generally, as the older botanists taught. However, we will leave this theoretical question for the present; we shall soon see that the fossil record has something to say on the point.

The character of the Angiosperms from which their name is taken is the closed ovary. The importance of this character lies chiefly in its effect on the mode of fertilisation. The ovary being closed, the pollen cannot get at the ovules directly, as it mostly does in the Gymnosperms or naked-seeded plants (Conifers, Cycads, etc.). Hence it is necessary for the pistil itself to arrange for the reception of the pollen. This is accomplished by the formation of the *stigma*, a downy or sticky surface, to which the pollen adheres when brought by insects, or blown by the wind. To bring the stigma into the best position for catching the pollen, it is often borne aloft on a long stalk—the style—formed between stigma and ovary. Thus great demands may be made on the powers of growth of the pollen, for each grain has to send out a tube long enough to reach all the way down the style and into the ovary, till its tip can enter an ovule. In a large Lily, for example, the distance which the pollen-tube has to cover may be quite six inches, a long way for so small a thing to have to grow. Thus the growth of pollen-tubes is a much more serious business in plants with closed ovaries than it is in those with exposed ovules, to which the pollen can be brought directly.

Ovaries, however, are important in other ways too. They protect the seeds while they are ripening (though the scales of a Gymnosperm-cone can do this also), and when they have developed into the fruit, they assist in dispersal in many different ways, sometimes by violently splitting open and acting like a catapult (as in Balsams), sometimes by attracting birds or other animals who eat the fruit and scatter the seeds (as in Gooseberries or Cherries), sometimes by developing hairy tufts, or parachutes so that the fruit can sail in the wind (as in Thistles). In these last cases and many others, the ovary or seed-vessel encloses only a single seed, and does not split open, but the whole fruit is sown. It is considered a bad mistake in botany to call the fruit of a Dandelion or a grain of Wheat a seed, but from the plant's point of view there is not much difference, for in such cases the ovary has come to be little more than an extra seed-coat.

However, what is important to us to note is that the ancestors of the Angiosperms, if we can ever find them, are likely to have had something approaching to a closed ovary.

There are other characters connected with the reproduction of Angiosperms quite as important as any we have already mentioned—I mean especially the changes which go on inside the ovule; these processes are such as to distinguish this class at once from all other plants. But at present we will not dwell on these points, because they do not happen to help us in interpreting that part of the fossil record which we are going on to consider. We may recall, however, that in Angiosperms a nutritive tissue, the endosperm, is developed, after fertilisation, in the embryo-sac, to supply food to the embryo. In a large proportion of Angiosperms, both Dicotyledons and Monocotyledons, this tissue is still present when the seed is ripe, and is only used up during germination; this is the case for example in Buttercups, Spurges, Grasses, and Palms. In another large group of families the endosperm is all, or nearly all, consumed while the seed is ripening, so that in the mature condition the embryo practically fills the whole seed: as in

Beans, Walnuts, Orchids and Water Plantains. The former are called albuminous seeds, the latter exalbuminous—old-fashioned terms based on a comparison between the endosperm of a seed and the white of an egg. Now, albuminous seeds are common to Gymnosperms and Angiosperms, though in Gymnosperms the development of the endosperm is absolutely different, for it is formed before fertilisation; exalbuminous seeds, however, are peculiar to Angiosperms and limited to certain families among them. The significance of this point will appear when we come to the fossils.

The reproductive characters of the Angiosperms are the most important. In the vegetative organs—stem, leaf and root—there is such a vast range of structure that we cannot profitably frame any short description common to the whole. One characteristic point is that the wood is, as a rule, more complicated in structure in Angiosperms than in Gymnosperms or any other plants. In particular, large continuous vessels, formed of rows of cells fused together end to end, are characteristic of the wood of Angiosperms, though vessels of a different kind occur in other groups of plants. The comparatively simple Gymnospermous wood, consisting essentially of tracheides (long, pitted, water-conducting cells with pointed ends) and medullary rays, usually serves to distinguish that class from Dicotyledons, though there are exceptions on both sides.

We have said very little so far about the distinctions between the two great classes, Dicotyledons and Monocotyledons, into which Angiosperms are divided, for our object has been to dwell on the points common to the whole group. Dicotyledons and Monocotyledons are as a rule, though with many exceptions, very different in habit, that is to say in general appearance; there is not often any difficulty in telling, at a glance, to which class a plant belongs. The outward distinction depends partly on the character of the leaves and partly on the mode of growth. It is generally said that in Dicotyledons the veins of the leaf form a network, while in Monocotyledons they are parallel. This is not strictly true, for in Monocotyledons also there is generally a kind of network of veins, consisting of a number of large longitudinal strands joined at intervals by more delicate transverse connections. The former are much the more conspicuous and form the "parallel venation" which is all that the casual observer notices. In most Dicotyledonous leaves the network is more complicated and more noticeable, the bundles branching and re-uniting in every direction. The different character of the foliage in the two classes depends chiefly on the venation, though the distinction is by no means constant.

The mode of growth of the stem, at least in the case of trees and other large plants, is very different as a rule, for the stem of a Dicotyledon can go on growing in thickness indefinitely, as we see in all our native trees and shrubs, and to a less extent in most other members of the class; in the great majority of Monocotyledons, when once the stem is formed, it is incapable of any further growth in diameter. Thus the stem and branches of a Dicotyledonous tree are thickest at the base, where growth has gone on longest, and the general form of the trunk is more or less conical, tapering from below upwards. In most Monocotyledons the stem is cylindrical, and towards the base may even have the form of an *inverted* cone.

These external differences go together with a great difference in the internal structure. In Dicotyledons the vascular strands in the young stem are arranged, as a rule, in a ring, with the wood on the inside and the bast on the outside; between the two lies the cambium, a delicate, actively growing tissue which goes on adding new wood internally and new bast externally as long as the plant lives, of course resting during the winter in plants, such as trees and shrubs, which live for a number of years. The wood, it will be remembered, serves chiefly, though not solely, to conduct the water absorbed by the roots up to the leaves, while the great function of the bast is to convey the food substances formed in the leaves to the places where they are wanted, either for immediate use in growth, or to be stored up for future purposes. In a Monocotyledon the vascular strands, instead of being ranged in a regular ring, are usually much more scattered, forming a number of irregular concentric circles, and often extending quite to the centre of the stem. Each has its wood and bast, as in a Dicotyledon, but as a rule, there is no cambium, so that no further increase can take place when once the strand has been formed. Further, the bundles remain separate, while in a Dicotyledon they are usually joined up, so as to form an uninterrupted zone of wood and bast. In all these respects the Dicotyledons resemble the Gymnosperms, while the Monocotyledons are quite peculiar and unlike any other plants. At one time it was thought that the Monocotyledons were the more primitive class of plants, but it is a curious fact that none of the older fossil plants (*i. e.* those before the Cretaceous epoch) have stems like Monocotyledons, while a very great number, like the living Gymnosperms, are of the same type with the Dicotyledons.

Another distinction between the two classes lies in the root-system. In a large proportion of Dicotyledons there is what is called a "tap-root," that is to say, a main root, forming the downward continuation of the stem; this tap-root lasts as long as the stem, and goes on branching, thus supplying all the new roots that are required. In Monocotyledons, there is usually a main root to start with, while the plant is still a seedling, but it soon dies away, and all the subsequent roots of the plant spring from the lower part of the stem. This character is correlated with the absence of any growth in thickness, a disability which the roots of Monocotyledons share with the stem. The original root of the seedling is too small to serve for the older plant, and as it cannot grow any stouter it has to be replaced by successively larger roots, which the stem sends out as it needs them.

There is no essential distinction between the *flowers* of the two classes, except the curious numerical one (subject, of course, to various exceptions) that the parts of the flower in Monocotyledons are usually in threes, while in Dicotyledons they are usually in fours or fives. In the petrified Cretaceous flower from Japan, referred to above, it was chiefly the fact that there were *three* carpels which suggested its being a Monocotyledon.

The most constant distinction, however, is that from which the names of the two classes are taken—the number of cotyledons in the embryo-plant.

The cotyledons are commonly called the "seed-leaves," though they by no means always perform the functions of ordinary leaves. They often serve only as storehouses of starch and other reserves of food, as in the Broad Bean, or as suckers to absorb food from the endosperm, as in Wheat and other Grasses. But in very many plants (as in the Wallflower, the Beech, the Sycomore, and the Lily) they ultimately expand and become green, resembling true leaves; this is probably their real nature in all cases. In Monocotyledons there is very constantly a single seed-leaf only; in Dicotyledons there are nearly always two, though there are a few exceptions. The difference is important, affecting the whole structure and mode of development of the embryo. It is remarkable that all Gymnosperms have essentially dicotyledonous embryos, for although in some of them, as in our common Fir-trees, there are many more than two cotyledons, it has been shown that the larger number arises by the early division of two primary seed-leaves.

Thus, in two characters of the greatest importance, the structure of the stem and the structure of the embryo, the Gymnosperms agree with the Dicotyledons and not with the Monocotyledons. It was on these grounds that the older botanists used to include the Gymnosperms under Dicotyledons, an arrangement which may still be met with in Floras and systematic works. This cannot, of course, be maintained now; the Gymnosperms now living are totally distinct from all Angiosperms in the whole organisation and development of the reproductive organs; it is, however, important to remember that they really have much more in common with Dicotyledons than with Monocotyledons.

What was the origin of the Angiosperms, the great sub-Kingdom which now dominates the vegetation of the world? This is the first question which must occur to our minds in approaching the subject of the Evolution of Plants. If an answer is to be found, we must seek it in the palæontological record, for an attempt to solve the problem by the mere comparison of living forms would only involve us in endless controversies and hypotheses; we will, therefore, in the next chapter, inquire what was the nature of the vegetation of the World in the age immediately preceding the first recorded appearance of the Angiosperms, and see whether among the plants which have come down to us from that period there were any which show signs of kinship with the Flowering Plants that succeeded them.

CHAPTER III

THE EVOLUTION OF THE FLOWERING PLANTS—THE EVIDENCE

THE Secondary period has been called "the Age of Gymnosperms," a description which applies perfectly to the greater part of the period, namely, from its beginning to the Cretaceous epoch, in which, as we have seen, the Angiosperms made their appearance. During the immense time represented by the deposition of the Triassic and Jurassic rocks the Gymnosperms were dominant in the Flora of the world, just as the Angiosperms have been in later periods, down to our own. The Age of Gymnosperms extended into the Cretaceous epoch, for it is only at a few places that Angiosperms have been found in Lower Cretaceous rocks. When one speaks of the Mesozoic or Secondary Flora, without further explanation, it is this earlier Flora which is meant, and not the later Angiospermous vegetation.

The characteristic Mesozoic Flora became fully established during the early part of the Triassic epoch, and from then onwards the vegetation maintained a remarkably uniform character up to the early part of the Cretaceous era.

During all this time Ferns were numerous, and a few Club-mosses and Horsetails have been found, but the greater and more prominent part of the land vegetation was Gymnospermous.

The Gymnosperms belonged to three main groups—the Conifers, the Ginkgoales (Maidenhair-trees), and the Cycads. We will dismiss the first two families in a few words, for, important as they are in themselves, their fossil members have no bearing on our immediate problem.

The Conifers were very well represented in Mesozoic times; remains attributed to all the living families—Araucarias, Pines, Cypresses, Yews, etc.—have been discovered in various Secondary rocks. The material is abundant and varied, though often difficult of determination. On the whole, however, the Secondary Conifers at present known were not so very different from those of our own day, and certainly there is nothing to suggest any relationship to the higher Flowering Plants.

The curious Maidenhair-tree (*Ginkgo biloba*), so named because its leaves closely resemble the leaflets of the Maidenhair Fern (*Adiantum*), is now a perfectly isolated species, without relations in the living Flora. Geological evidence shows that it is the last survivor of an ancient family, which flourished during Secondary times, and can even be traced back to the Primary rocks. The race might not have lasted down to our times at all, if it were not that the Maidenhair-tree was regarded as sacred in China and Japan and grown in the gardens of the Buddhist temples. There is some doubt whether it is known except in cultivation, though probably it may be wild on the mountains of Western China.

The Maidenhair-tree, in its plum-like seeds and mode of fertilisation, has much in common with the Cycads, the great feature of Mesozoic vegetation. As it is this class of plants which has come to be of chief interest in considering the question of the evolution of Angiosperms, we must discuss them rather fully, especially as they are not very familiar plants except to botanists.

The Cycads have not been so much diminished in the course of ages as the family of the Maidenhair-tree, though, in proportion, perhaps, the reduction is not much less. They are represented at present by a small family, the Cycadaceæ, which includes nine genera and about 100 species. They are handsome plants, with foliage not unlike that of some Palms, whence their popular name, Sago-palms, for a kind of sago is obtained from some of them. With the true Palms, however, they have nothing whatever to do. Some few are often met with in cultivation under glass, but it is only in botanic gardens that they are to be seen in any numbers. There is a magnificent collection at Kew, chiefly in the Palm-house. They are natives of the tropical

Fig. 1.—Male plant of *Stangeria schizodon*, bearing several cones. Kew.

and sub-tropical regions of the earth. Of the nine genera, one, the type-genus *Cycas* (fig. 6,) is common to Asia and Australia; two, *Macrozamia* (figs. 2 and 3) and *Bowenia*, are Australian; two, *Encephalartos* and *Stangeria* (fig. 1), are South African; while four, *Zamia*, *Ceratozamia*,

Dioon, and *Microcycas* (fig. 4), are American, the last-named, containing a single species, having only been found in the island of Cuba. The family is thus widely distributed, though its members are not often met with in great abundance.

In a few cases the living Cycads attain the stature of trees. *Microcycas* (in spite of its name, which means little Cycad) grows to a height of 30 feet; a species of *Dioon*, from Mexico, attains 40 feet, and an Australian species of *Cycas* is said to reach as much as 60 feet in height. Some species of *Encephalartos*, when old, have the character of low trees. In most species of the family, however, the stems, though long-lived and often massive, remain short.

In nearly all cases the leaves, which are always of considerable size, are simply pinnate, the midrib bearing two rows of leaflets; in the Australian *Bowenia* the leaflets are again divided in the same way, and in a species of *Cycas* recently discovered in Annam, each leaflet is forked twice or more. The leaves of the African *Stangeria* (fig. 1) are veined exactly like those of a Fern, so that the plant was described as a Fern of the genus *Lomaria* when first brought to England, and its true nature was only revealed when its cones appeared. Another Fern-like feature in some Cycads is the way the leaves are coiled up in the bud. The crozier-like (*circinate*) form of the young Fern-frond before it unfolds will be familiar to all readers. In *Zamia* the whole

Fig. 2.—*Macrozamia Moorei*, stem. Kew.

leaf resembles a Fern-frond in this respect, while in *Cycas* the leaflets are circinately coiled. As we shall see presently these little points of resemblance to Ferns are significant.

In most Cycads each leaflet is traversed by a number of more or less parallel veins; in *Stangeria* there is a well-marked midrib, and the veins which pass off from it on either side are forked like those of a Fern-frond; in *Cycas* itself, each leaflet has a midrib only, with no lateral veins at all.

In many Cycads the stem is clothed with an armour, made up of the bases of the old leaves, which continues to protect the trunk for many years, as in *Encephalartos*, *Dioon*, *Ceratozamia*, *Macrozamia* (fig. 2), and most species of *Cycas*. In the rest of the family the armour is not so persistent. In most species scale-leaves are produced at intervals, and the armour is then composed of alternating zones of the larger and smaller leaf-bases, corresponding respectively to the green leaves and the scale-leaves.

The stem has a very large pith, the diameter of which may be half that of the stem as a whole; in this respect the Cycads differ conspicuously from Conifers and from most Dicotyledons. Otherwise the structure of the stem, at least in its simplest form, is of the usual Gymnospermous type, which, as we have seen, is common to Dicotyledons also. The zone of wood and bast is often poorly developed, the stem having a succulent character; thus in a specimen of *Dioon edule*, a Mexican species, the stem, though 120 years old, had a vascular zone only half a centimetre (one-fifth of an inch) thick, about one-fifteenth of the radius of the stem. Where the stem increases much in diameter it often does so in an unusual way, the original cambium ceasing to be active, while a new cambium appears further to the outside; this may take place several times. Thus the wood comes to be built up of a number of successive layers, each produced by its own cambium; this is the case in *Cycas*, *Macrozamia*, and other genera. There are many other points of interest in the anatomical structure, into which we will not enter now; we will go on at once to the organs of reproduction.

In the Cycads now living the sexes are always separate, on different plants. Generally speaking the fructifications are in the form of cones, comparable to those we are familiar with in the Coniferæ, but simpler. This holds good for the male fructification all through the family (figs. 1 and 3); in eight out of the nine genera the female fructifications are also cones (fig. 4), but in *Cycas* itself a different arrangement prevails (fig. 6). The cones are stalked structures, borne at the top of the stem; in fact, it appears that in many cases, at any rate, the cone is really developed from the apex of the whole plant, though pushed aside by the subsequent growth of the stem.

The two kinds of cone are much alike in outward appearance, the female being generally the more robust.

Fig. 3.—*Macrozamia spiralis;* male plant, bearing several cones. Lower leaves removed. Kew.

We will begin with the male fructifications, which are among the largest known, the cone itself, without the stalk, often reaching a foot or more in height (fig. 3).

The cone consists of an axis bearing great numbers (up to 600) of closely packed scales, which in their turn bear the pollen-sacs. These scales are therefore equivalent to the stamens of Flowering Plants, but they are widely different from them in structure. The form of the scale

Fig. 4.—*Microcycas calocoma;* female plant, bearing a cone. Cuba. After Caldwell.

is often shield-like, with an enlarged end; sometimes it is more like a simple leaf in shape. The pollen-sacs are borne on the lower surface, in large numbers, ranging from a dozen or so in *Zamia,* to something like a thousand in *Cycas.* They are in the same position as the sporangia or spore-sacs on the fronds of Ferns, and like them, they are arranged in little groups or sori (2–6 sacs in each group), though this is not easily made out when the stamen is mature. Inside each sac a great number of pollen-grains are produced, the sac opening by a slit to let them escape, while the scales of the cone separate a little to give the pollen a free passage into the open air. The whole arrangement is quite different from that in the stamen of an Angiosperm, where there are usually four long pollen-sacs, ranged two and two on each side of the anther; it is more like what we find in Conifers, though in that family the pollen-sacs are nothing like so numerous.

In describing the female cones, we must first of all leave *Cycas* out of consideration; what follows refers only to the other eight genera.

The scales of the female cone are larger and less numerous than those of the male, but very similar to them in form (see figs. 4 and 5, *Microcycas*); those of *Dioon* are the most leaf-like in character. These scales may be called carpels, for they bear the ovules, but, as in the case of the stamens, they have little in common with the corresponding organs of the Flowering Plants. Each carpel regularly bears two ovules, which are seated on the edge of the expanded portion. The ovules are very large, and may grow to the size of a small plum even without being fertilized. (Those of *Cycas* are still bigger.) The cones, when ripe, are often beautiful objects; in *Enceph-*

alartos, for example, the scales of the cone are orange-coloured, and the seeds, which show between them, scarlet. It is something like the colouring of the Spindle-berry, but on a huge scale, and with the colours of seed and carpel reversed.

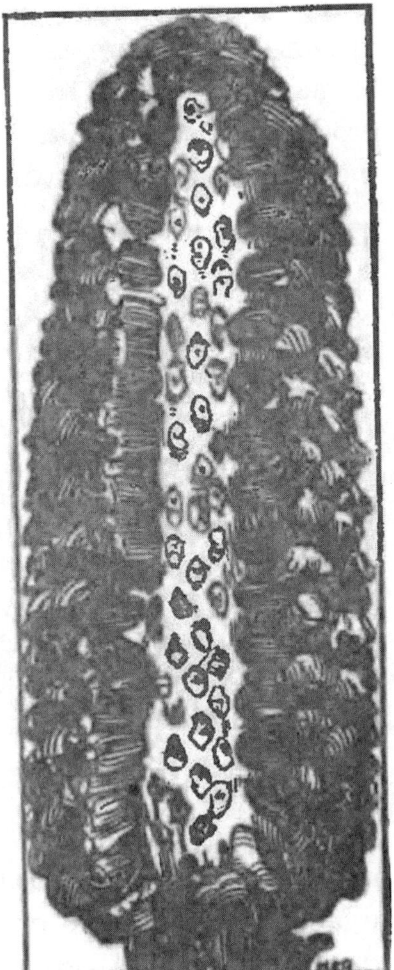

Fig. 5.—*Microcycas calocoma;* female cone, cut open to show the ovules. After Caldwell.

So far all the Cycads we have described have been cone-bearing plants, their fructifications showing a general resemblance to those of Coniferæ, though very different in detail, and on the whole more simply organised. Though quite unlike the true flowers of the Angiosperms, the cones have still much in common with flowers, for cones, as well as typical flowers, are branches specially set apart to bear carpels or stamens. Now, with one exception, all living plants which bear seeds have flowers or cones; *i. e.* there are always special branches formed, on which the leaves concerned in reproduction are borne. The one exception is the female plant of *Cycas*. Here no separate cone

Fig. 6.—*Cycas revoluta;* centre of female plant, showing some leaves and the rosette of leaf-like carpels, bearing seeds. Kew.

or flower is formed; the carpels are leaf-like organs borne directly on the main stem of the plant, just as the ordinary leaves are (fig. 6). When the flowering season comes, the carpels grow out in a rosette, all round the stem, and when their work is done they drop off, and are succeeded by new green leaves, as before. Hence the armour of the stem of a female *Cycas* is composed of the bases of three different kinds of leaf — the foliage-leaves, the scale-leaves, and the carpels. In thus bearing its reproductive leaves or *sporophylls* on the main stem of the plant, a female *Cycas* is on the same level as a Fern, such as the Ostrich Fern, in which the sporophylls are distinct from the ordinary fronds but borne on the same stem. No other living Seed-plant is so simple as this, though the same arrangement was common among the Seed-plants of Palæozoic age (see Chapter IV).

The carpels of *Cycas* are in themselves remarkable structures, and show their leaf-nature more clearly than those of any other plant. In *Cycas revoluta*, for example, the species most commonly cultivated, the carpel, about nine inches in length, has a broad, deeply-divided, hairy blade (fig. 6); the ovules, usually six in number, are borne on either side of the lower, stalk-like portion. In *C. circinalis* the carpel is narrower, and the divisions of the blade reduced to teeth; in an Australian species, *C. Normanbyana*, there are only two ovules. The species of *Cycas* are more gymnospermous than most Gymnosperms, for the ovules are fully exposed all through, until they are ultimately shed as the ripe seeds. The seeds of *Cycas* reach an extraordinary size — in some species they are about as big as a peach; they also resemble a peach or plum in structure, for the outer part is soft and fleshy, and encloses a stone. The whole organ, however, is here an excessively developed *seed*, whereas in the peach everything except the kernel inside the stone belongs to the *fruit*. In some species of *Cycas* there is the same brilliant contrast of colour which we mentioned in *Encephalartos*. In *Cycas revoluta* the bright-red ovules or seeds on their yellow carpels have a very gay appearance.

We see then that the Cycads as a family are decidedly simpler in their reproductive arrangements than other Seed-plants. *Cycas* in particular, as regards the female plant, stands quite alone, as the only living genus of seed-bearing plants in which neither flower nor cone is developed; the carpets are not only obviously leaf-like in form, but are borne, among the ordinary green leaves, on the main stem of the plant.

But there are other points in which the Cycads even more clearly betray a primitive character among Seed-plants, approaching the Spore-plants. This is most strikingly shown in their mode of fertilisation, discovered fourteen years ago by two Japanese botanists—Hirase and Ikeno—and inpendently just afterwards by Webber in America. Like most of the Cryptogams, but unlike any other living Seed-plant except the Maidenhair-tree, the Cycads effect their fertilisation by means of actively moving male cells, the *spermatozoids*.

This important discovery was first made in the genus *Cycas*, by one of the Japanese observers just mentioned, and was immediately afterwards extended to *Zamia* by Webber's researches; spermatozoids have since been demonstrated in *Stangeria, Ceratozamia, Microcycas* and *Dioon*, and are no doubt common to the whole family. It may be added that the very first Seed-plant in which spermatozoids were found was not a Cycad, but the Maidenhair-tree (*Ginkgo*), where they were observed by Hirase a month or two before their discovery in *Cycas* by Ikeno. A short account of the process as it has been observed in *Cycas* and *Zamia* may be given before we further consider the significance of the facts.

The ovule, at the time when it is ready to receive the pollen, is about the size of a grain of Maize, or a small Hazel-nut. It consists of an outer envelope and a central body, the two being closely united except at the top, where a narrow passage (the micropyle) is left open in the envelope, leading down to the central body, or *nucellus*, as it is technically called (fig. 7, D). The apex of the nucellus becomes excavated to form a deep pit, the pollen-chamber, a character ony shown by the Cycads and the Maidenhair-tree among living plants, though common in Palæozoic seeds (see Chapter IV).

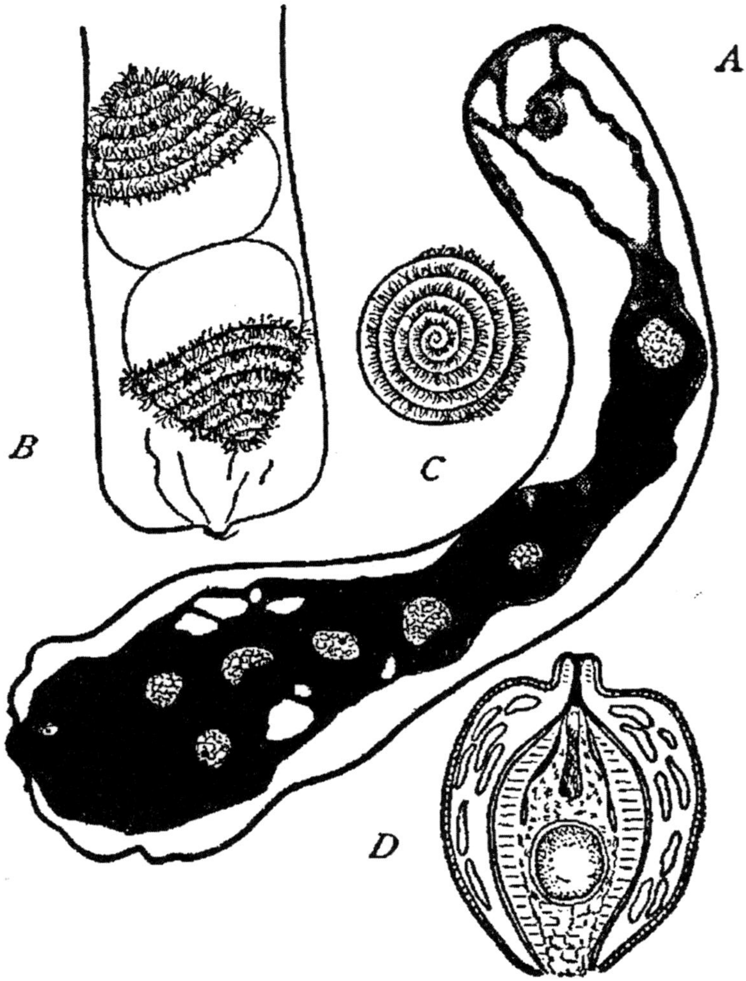

Fig. 7.—(A) Pollen-tube of *Microcycas*, containing numerous male cells × 90; (B) end of pollen-tube of *Zamia*, containing the two ciliated spermatozoids; (C) a free spermatozoid, showing the spiral band bearing the cilia (B and C × 80); (D) longitudinal section of ovule of *Cycas*, showing embryo-sac, nucellus with pollen-chamber, and integument slightly magnified.

A after Caldwell; B and C after Webber; D after Griffith.

The pollen, whether blown by the wind, or, as seems likely to be the case in some South African Cycads, conveyed by insects, is received in the micropyle by a drop of gummy substance, in which the pollen-grains stick. As the drop evaporates or is re-absorbed, the pollen-grains are drawn down through the narrow passage into the pollen-chamber below (fig. 7, D). Here each grain anchors itself by sending out a branched tube into the surrounding tissue. Thus *pollination* is effected; the actual fertilisation does not take place for some months later, when the ovule has grown to something like full size and become, to all appearance, a seed. In the interval between pollination and fertilisation, the embryo-sac or megaspore, embedded in the central body of the seed, has grown to enormous dimensions, filled itself with tissue (the endosperm or prothallus) and developed several egg-cells at its upper end. It will be remembered that it is characterstic of Gymnosperms for the endosperm to be formed *before* fertilisation. In Cycads, everything is on a great scale; even the egg-cells are quite large enough to be seen with the naked eye. Meantime the pollen-tube has also grown much, and, besides other changes, has developed in its interior two large cells, the male-cells or sperms. Each of these has a long spiral band round one end, on which a multitude of delicate filaments or *cilia* are developed; these are the organs of motion (fig. 7, B and C). The pollen-tube, distended with water, now grows a little, and brings the end where the sperms are, near the egg-cells; the sperms begin to rotote while still in the tube and now show themselves to be true, active spermatozoids. The pollen-tube bursts; the spermatozoids are set free, and by means of their actively vibrating cilia, swim the short distance necessary, through the water discharged from the pollen-tubes, reach the egg-cells, unite with them, and so effect fertilisation. Each spermatozoid can fertilise one egg-cell and one only.

The Cycads and the Maidenhair-tree are the highest plants which have retained the Cryptogamic method of fertilisation by actively moving cells; it will be remembered that spermatozoids are the agents of fertilisation in all Ferns, Mosses and their allies, and in a large number of Seaweeds and other Algæ. In animals, too, the male cells have the power of active movement. The higher plants rise in the scale of evolution, the more do they tend to lose their animal characters; some of the lowest Algæ have the power of movement through most of their lives; in higher forms it is only the spore-stage that keeps this power; in still higher plants, like Mosses and Ferns, the spores too have lost their mobility, but the male cells still keep it, as a means of reaching the egg-cells. Lastly, this particular kind of movement is lost altogether; the more highly developed a plant becomes, the less, generally speaking, is it like an animal.

The Cycadean method of fertilisation is particularly interesting because it is exactly intermediate between the purely Cryptogamic process in which everything depends on the active movement of the spermatozoids, and the process in the higher Seed-plants, in which the growth of the pollen-tube is all important, the sperms being carried as mere passengers to their destination. In Cyads a pollen-tube is formed, and its growth is so directed as to help the spermatozoids on their way, but the final stage of their journey to the egg-cell still has to be accomplished by their own efforts.

It is interesting to recall that the great German botanist Hofmeister, fully forty years before the discoveries of Hirase and Ikeno, came very near to predicting their results, for he said that "impregnation in the Coniferæ takes place by a pollen-tube in the interior of which spermatozoa are probably formed." Though this is not the case in the Coniferæ generally, it has proved to be true in *Ginkgo* (which was regarded as a Conifer in Hofmeister's time), as well as in the allied Cycads.

In most Cycads only two spermatozoids are

produced in each pollen-tube; there is, however, a striking exception in *Microcycas*, a most remarkable plant of which scarcely anything was known till it was rediscovered three years ago, by the American botanist Caldwell, on the mountains of Cuba; instead of being a little plant as was supposed before, *Microcycas* turned out to be a tree thirty feet high, with female cones a yard long (figs. 4 and 5). The interesting point in the reproduction is that each pollen-tube produces no less than sixteen spermatozoids (fig. 7, A)—a fact which offers an interesting parallel to the conditions in some of the Palæozoic Seed-plants (Chapter IV). In *Microcycas* the number of egg-cells is extraordinarily large, as many as two hundred in a single ovule, whereas from three to five is about the usual number in other Cycads.

After fertilisation more than one embryo may go on developing for a time; in *Microcycas* there may be a dozen or more. Sooner or later one gains the upper hand, and is alone present by the time the seed is ripe. The embryo is then embedded in the endosperm, which always fills the stone of the seed and is only used up when germination takes place. An interesting point in Cycads is that the embryo may scarcely develop at all till the seed is sown, so that the apparently ripe seed may contain no obvious embryo. It is common in Cycads for there to be no resting-stage in the seed, the development of the embryo going straight on into that of the seedling. There is much diversity, however, in this respect.

The embryo usually has two cotyledons, which are often fused together. In *Ceratozamia* one of the cotyledons is, as a rule, abortive. On germination they remain in the seed, serving as suckers to absorb food from the endosperm. It will be remembered that in most Conifers (though not in all) the cotyledons expand as green leaves.

I have dwelt at some length on the modern Cycads because a knowledge of this family is essential, not only in order to understand the Mesozoic Flora, which immediately preceded the advent of the Flowering Plants, but also for comparison with the Palæozoic Seed-plants (see Chapter IV). We will now see what the Cycads were like in Mesozoic times, and inquire how far they may help us towards a solution of the great problem of the evolution of the Flowering Plants.

At the present day the Cycads scarcely amount to one in a thousand of the vascular plants; in the Secondary Floras about one plant in every three was a Cycad. The little family now living represents what was then the dominant class of plants. But it must be understood from the outset that in speaking of Mesozoic Cycads we use the name for a far wider group than the surviving family; the whole class may bear the name of Cycadophyta, of which only the one family, Cycadaceæ, has come down to our own times.

The Cycadophytes of the Mesozoic had a world-wide distribution, extending from the tropics all through the temperate zones into the Arctic regions of Greenland and Siberia, and the Antarctic of Louis Philippe Land.

They were abundant in our own country and their stems are familiar objects in the quarries of Portland, where they are called "Crows' Nests" by the workmen. The vegetation of the world in those days was astonishingly uniform; quite similar forms of Cycadophyta have been found in Yorkshire, Mexico, India, and even in regions where only the hardiest Arctic vegetation can grow at all in these days.

The living Cycads, as we have seen, are apparently a simple group of Flowering Plants; they show, in various features, especially in their mode of fertilisation, in the arrangement of their pollen-sacs, and in the carpels of *Cycas*, a decided approach to Cryptogams. Among the Mesozoic Cycadophytes it is rare to meet with such simple forms. In a few cases fructifications similar to those of recent Cycadaceæ have been found; in particular, the genus *Cycas* has been recognised, by means of its characteristic carpels, as far back as the Lias. Leaves closely resembling those of recent genera such as *Cycas*, *Zamia* and *Dioon* are frequent, but such evidence is not to be relied on, for in certain cases these leaves have proved to belong to plants widely different from the modern Cycadaceæ, though no doubt belonging to the class Cycadophyta. So far as our present knowledge goes, the family Cycadaceæ, though existing in Mesozoic times, seems to have attained no great development.

The vast majority of the Mesozoic Cycadophyta hitherto investigated have proved to belong to a group resembling the recent Cycads in habit and vegetative structure, but beyond comparison more highly organised as regards the reproductive organs. This is the family of the Bennettiteæ, which are well represented in Britain, France, Italy, and other parts of Europe, but far more abundantly in the United States, where Maryland, Dakota and Wyoming are richer in these fossils than any other part of the world.

In Britain, Cycads of this family have been found in the Lower Greensand and the Wealden of the Isle of Wight, in the Wealden of Sussex, in the Purbeck Beds (Upper Jurassic) of East Dorsetshire, in the Great Oolite of Brora in the north of Scotland, and in the Inferior Oolite of the Yorkshire coast. The most valuable specimens are those which are petrified, so that the internal structure is preserved; one of the most famous fossils of this kind is *Bennettites Gibsonianus*, from Luccombe Chine in the Isle of Wight; in this species the structure of the fruit was first revealed by the work of Carruthers and Solms-Laubach. Other European specimens have contributed much to our knowledge, but the most important results of all have come from the investigation of the rich American material, during the last ten years or so, by Dr. Wieland of Yale. To him we owe our first knowledge of the structure of the *flowers*, a word which, as we shall see, may appropriately be used for the fructifications of these plants. About sixty species have been distinguished from the American Upper Jurassic and Lower Cretaceous rocks; the individual specimens are extraordinarily abundant; from the Black Hills of Dakota alone, nearly a thousand silicified trunks, referred to twenty-nine species, have been obtained.

In general aspect the Bennettiteæ resemble the shorter-stemmed Cycads of the present day; none are known to have attained any very great height; the leaves were scarcely different from those of such recent Cycads as *Zamia*. The stem was clothed with an armour of leaf-bases, but there is one peculiarity which at once distinguishes these fossil trunks from those of any living Cycads. In the Bennettiteæ we find, scattered among the bases of the leaves, the scars of leafy branches resembling large buds; these are the remains of the fructifications, which in these plants were borne laterally and in great numbers, instead of growing

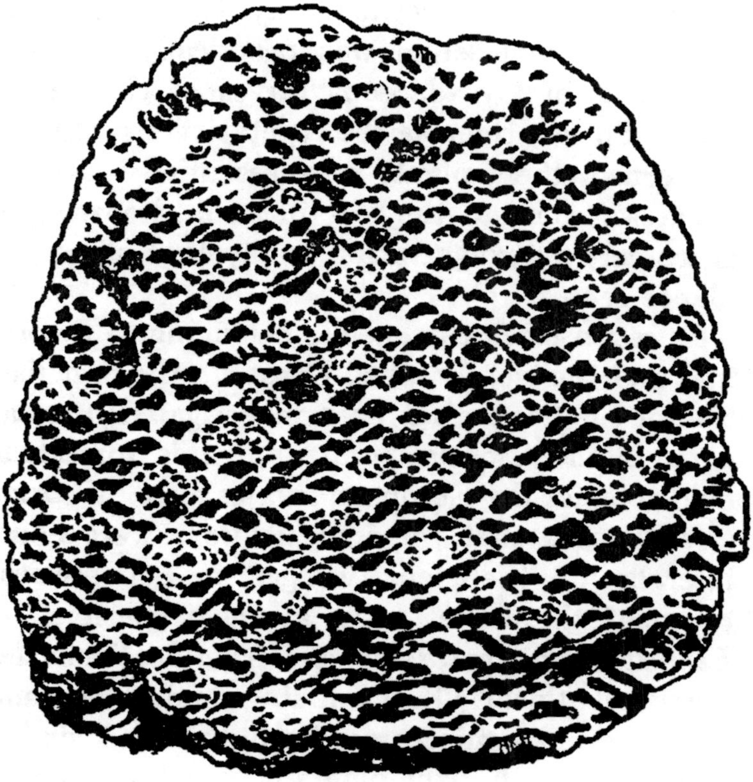

Fig. 8.—*Bennettites marylandicus.* The earliest described American fossil Cycad, showing the leaf-bases and fructifications. After Wieland.

out, a few at a time, from the top of the stem, as in the recent family (fig. 8).

The anatomy of the stem need not detain us; it was essentially the same as in the simpler living Cycads, that is to say, there was a ring of wood and bast surrounding a wide pith. The vascular strands supplying the leaves took a more direct course than in the recent family, the structure of the leaves themselves was practically identical with that of Cycadaceous leaves of the *Zamia* type. So far, then, as the vegetative structure is concerned, these plants were essentially Cycads; when we come to the fructification, however, it is scarcely an exaggeration to say that all resemblance ceases.

In seven American species belonging to *Bennettites* and the closely allied *Cycadella* the organisation of the flowers has been investigated; they all agree in the essential points, and this appears to hold good for the European species also. We have already seen that the fructifications were borne laterally on the stem, wedged in between the bases of the leaves. On a single specimen of *Bennettites dacotensis* sixty-one fructifications were found, all at nearly the same stage of development; Dr. Wieland conjectures, from this and similar observations, that the plants flowered once for all and then died, like many Palms and Bamboos at the present day.

The species just mentioned, *Bennettites dacotensis*, may serve as a type. The whole flower is here about twelve centimetres (nearly five inches) long, and protrudes a little beyond the leaf-bases by which it is surrounded (fig. 9). The lower half of the flower consists of a stout axis bearing numerous bracts, which are spirally arranged and enclose the essential organs of reproduction.

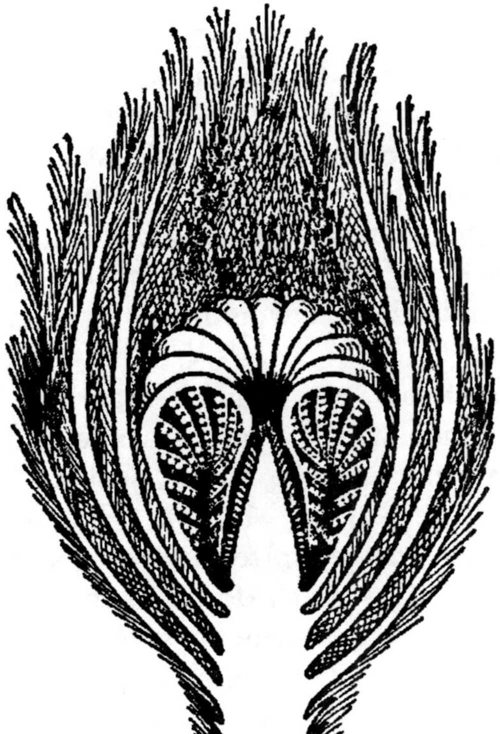

Fig. 9.—*Bennettites*. Restoration of unexpanded flower, showing the hairy bracts, the infolded, compound stamens, and the conical receptacle, bearing the ovules. After Wieland.

Within the uppermost bracts are the stamens, which are not in spiral order, but form a ring or whorl, like those of most Angiosperms. In this species the number of stamens is about eighteen or twenty; they are very large and complicated organs, quite different from the stamens of any Angiosperm or any other Gymnosperm. Each stamen, if straightened out, would be about ten centimetres (four inches) long; as a matter of fact, however, they are always found folded inwards, as if the flower were still at the stage of a bud (fig. 9). Each stamen is a pinnate leaf, with about twenty pairs of leaflets, bearing the pollen-sacs, which are ranged in two rows on each leaflet; the longer leaflets bear as many as twenty pollen-sacs each.

The stamens are joined together by their stalks, which thus form a tube surrounding the female part of the flower, just as is the case in the flower of a Mallow. Higher up, the stamens become separate, and are bent inwards, so that their tips reach down to their stalks.

The pollen-sacs, borne in such large numbers on the stamens, are themselves compound structures, for each sac is divided into numerous compartments, in which the pollen was produced. Such compound pollen-sacs are almost identical in structure with the "synangia," or compound spore-sacs of certain Ferns belonging to a small tropical family, the Marattiaceæ (see fig. 17, p. 150). Indeed the stamens of Bennettiteæ were altogether much more like the fertile fronds of a

Fern, than the stamens of any other plant. Each pollen-sac was attached to its leaflet by a short stalk.

The centre of the flower, just as in an Angiosperm, is occupied by the female apparatus or *gynæceum* (it is better to avoid the use of the word *pistil* for reasons which will appear presently). This consists of a conical axis, the receptacle, on the sides of which a great number of stalked ovules are borne, intermingled with barren scales. The structure of the gynæceum is better made out in the mature specimens, where the flower has become a fruit. The point to note here is that the arrangement of the organs in the fructification of the Mesozoic *Bennettites* is just the same as in a typical Angiospermous flower—on the outside the bracts, which might quite well be called a perianth, then a ring of stamens, and finally the female apparatus in the middle. This arrangement seems to have been general in *Bennettites* and to have extended to some of its allies, though in other related plants the flowers are said to have been of separate sexes.

In the more mature fructifications the stamens have disappeared, leaving only a rim round the base of the receptacle, to show where they were inserted. One of the best examples of the fruiting stage is found in *Bennettites Gibsonianus*, the Luccombe Chine fossil; this plant was the first to be discovered in a fertile condition, and from this discovery our knowledge of the Bennettiteæ may be said to date, for though related plants had been discovered much earlier, their structure was not understood.

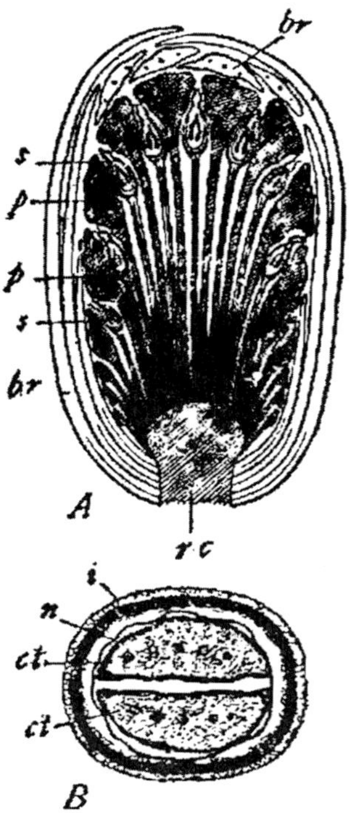

Fig. 10.—*Bennettites Gibsonianus.* (A) Diagram of fruit in longitudinal section: (*rc*) receptacle; (*br*) bracts; (*s*) stalked seeds, each containing a dicotyledonous embryo; (*p*) sterile scales, with dilated ends forming the pericarp. (B) transverse section of a seed; (*i*) seed-coat; (*n*) remains of nucellus; (*ct*) the two cotyledons of the embryo, cut across. × about 10. From Scott, Studies.

The fructifications are borne in exactly the same way as those of the American species, and their structure is no doubt essentially the same, only they happen to have been fossilised at a later stage of development, when the flower had become a fruit.

The bracts still enclose the fruit, but the stamens have gone, and the gynæceum has enlarged so as to fill up all the space within the bracts (fig. 10, A). The receptacle (more rounded in this species than in the American one already described) bears a large number of seeds, each seated on a long slender stalk; the spaces between the seed-stalks are packed with sterile scales, which are enlarged at their ends and fit closely together, forming a kind of ovary-wall, or pericarp. This, however, is not completely closed, for wherever there is a seed a little gap in the wall is left, into which the open end or micropyle of the seed fits, and so reaches the surface. The seeds are beautifully preserved; their structure is simple compared with that of the seeds of living Cycads, for in the fossils they were well protected by the fruit in which they were embedded.

The most interesting fact about the seeds is that they contain well-preserved embryos. It is only in this group of plants that fossil embryos have been found; it is an extraordinary piece of good fortune that such delicate bodies should have been preserved at all.

The embryo is dicotyledonous (fig. 10, B); the two fleshy cotyledons are turned towards the base of the seed and the young root towards the micropyle, just as in ordinary seeds at the present time. The embryo practically fills up the whole interior of the seed; there was little or no endosperm when the seed was ripe. Thus the seeds were "exalbuminous," a state of things quite unknown among living Gymnosperms, and only found in a part of the Angiosperms. We must thus regard it as a very advanced character, and not at all what we should have expected to find in a Cycad.

There are, then, three distinct characters in which these Mesozoic Cycadophytes approached the Angiosperms:—

1. In the presence of flowers, organised on the same general plan as the typical flowers of Angiosperms.
2. In the formation of a fruit, enclosing the seeds.
3. In the exalbuminous nature of the seed.

The coincidence of these various characters can scarcely be without significance, and we seem justified in the conclusion that of all known plants the Bennettiteæ were the most nearly akin to the Angiosperms. Both from the structure of the embryo and that of the stem it is evident that the affinity was with the Dicotyledons rather than with the Monocotyledons.

There remain, however, a number of characters in which the fossil family differs strikingly from the Angiosperms; the principal points of difference are:—

1. The habit.
2. The complexity of the stamens.
3. The structure of the gynæceum.
4. The probable mode of fertilisation.

In habit, as we have seen, the species of *Bennettites* and its immediate allies were altogether Cycadean—there was no approach to the external

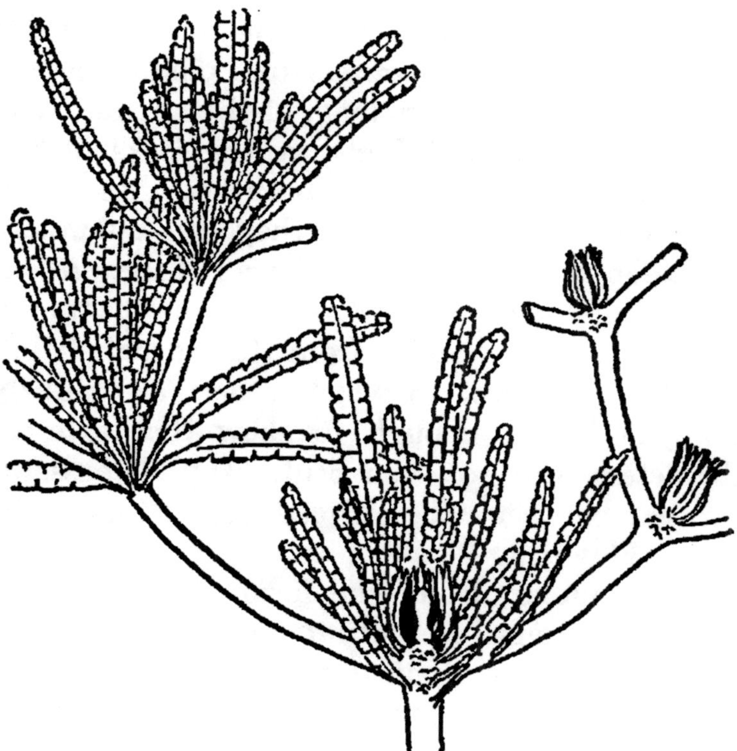

Fig. 11.—*Wielandiella angustifolia.* Restoration of part of plant, showing the forked stem, the leaves and the flowers. After Nathorst.

characters of any family of the higher Flowering Plants. But when we come to take a wider view of the group, we find that this Cycad-like habit was not constant. In the genus *Wielandiella* of Nathorst, derived from the Upper Trias of Sweden, the habit is quite unlike anything which we are accustomed to associate with Cycads (fig. 11). The long, slender stem was repeatedly forked, with rosettes of narrow, pinnate leaves only about three inches long, at the angles. The flowers were seated singly in the forks of the stem between the two branches. As Prof. Nathorst has lately shown, they had essentially the same structure as the flowers of *Bennettites*. Here, then, we have a member of the group which departs widely from the usual habit, and this instance, to which others are already being added, is enough to show that external characters do not constitute any impassable barrier between the Cycadophytes and the higher Flowering Plants.

The complex, frond-like form of the stamens and their immense output of pollen-sacs are striking features, indicating an affinity to the Ferns, and very unlike anything in Angiosperms. It is true that branched stamens occur in the latter, as in the Mallows and the Castor-oil plant, but this appears to be only a secondary and not a primitive character. In the great majority of Flowering Plants the stamen has a very constant structure, bearing four elongated pollen-sacs, arranged two and two on either side. It has been well suggested that each pair of pollen-sacs may correspond to a synangium of Bennettites, with its double row of compartments. In a few Angiosperms (*Mimosa*, for example), similar compartments exist. But have we any evidence of a transition from the complex Bennettitean stamen to the simple Angiospermous type? A fossil recently found by Dr. Wieland in the Upper Trias of the Mixteca Alta in Mexico helps to bridge the gap. These beds are extraordinarily rich in remains of Mesozoic Seed-plants, especially *Williamsonia*, a genus which has been known since the early part of the last century, when *Zamia gigas* (now *Williamsonia gigas*) was first discovered in the Yorkshire cliffs. The flowers of *Williamsonia* (unfortunately preserved only as casts) have been repeatedly investigated, and it has now been established that both male and female organs were essentially of the same type as in *Bennettites*. It appears, however, though this is not perhaps quite certain, that in *Williamsonia* there were distinct male and female flowers.

Now in Dr. Wieland's Mexican specimen, which he calls the El Consuelo *Williamsonia*, from the quarry where it was found, there is a ring of stamens, united below into a bell-shaped tube, only the upper parts being free. The point of interest is that the stamens instead of being pinnate are simple, the pollen-sacs are inserted in two rows, directly on the main stalk. This is a distinct step towards the simpler type of stamen, and it is interesting to note that the occurrence of such a stage was predicted theoretically, by Arber and Parkin, two years before its discovery in the Mexican fossil. It may also be mentioned that in *Wielandiella,* the plant referred to above in speaking of the external characters, the stamens, according to Prof. Nathorst, were comparatively simple structures.

It is curious that in some respects the older Cycadophytes (*Wielandiella* and the Mexican *Williamsonia*) seem to approach the Angiosperms more nearly than those from later Mesozoic strata. This may perhaps mean that the typical Bennettiteæ were a side-branch, a little further removed than their predecessors from the main-line of evolution of the Flowering Plants.

The peculiar structure of the gynæceum in all known Bennettiteæ separates them at present from the Angiosperms, and as yet no transitional forms have been discovered. The ovules, it will be remembered, are seated each on a long stalk, and between them are the sterile scales, which expand at their ends to form collectively the pericarp. The stalk on the end of which the ovule is borne is best regarded as a much simplified carpel; the barren scales may be simply comparable to other scale-leaves, such as those of recent Cycads, or may be carpels which have ceased to be fertile. On either view there is a great difference from Angiosperms, in which the ovules are enclosed by the carpels that bear them; in Bennettiteæ they are enclosed by a distinct set of organs, the sterile scales. It is a little difficult to imagine how the passage could have been made from the one arrangement to the other; the facts rather suggest that we have not yet got the true ancestors of Angiosperms, but only other branches of the same great stock.

As regards the fourth point, the mode of fertilisation, we have no direct knowledge, for no detailed description of the structure of the gynæceum at the *flower* stage has yet been published. We know, however, that at a later stage, when the seeds were ripe, they reached the surface of the fruit, their micropyles protruding a little through the narrow openings of the pericarp-wall. Probably this was the case throughout, and if so, the ovules were no doubt fertilised directly, the pollen-grain reaching the micropyle. In other words, the facts point to the Bennettiteæ having been still Gymnospermous as regards their mode of fertilisation. This, if it were so, is not surprising, for it is not suggested that the Bennettiteæ were Angiosperms, but only that they were near the probable line of Angiospermous descent.

It has already been pointed out that the nearest affinities of the Bennettiteæ were clearly with Dicotyledons. The particular family with which they seem to have most in common is that of the Magnoliaceæ, to which the stately Tulip-tree and many other beautiful trees and shrubs belong. In this order the numerous, free perianth-leaves, not usually distinguished into calyx and corolla, their frequent spiral arrangement, the hypogynous stamens (inserted directly on the receptacle below the gynæceum), the often elongated axis of the whole flower, and especially the long, cone-like receptacle, with its numerous carpels, all suggest comparison with the fossil forms. The resemblance extends in a greater or less degree to other related families, such as our familiar Ranunculaceæ, and the Nymphæaceæ (Water-lilies). It is interesting to recall that *Drimys* (Winter's Bark), a member of the Magnoliaceæ, and two or three plants of allied families, are the only Dicotyledons which are known to have wood of the same simple structure as the Gymnosperms. This group of families has been regarded as primitive by many botanists; there is evidence for the presence of Magnoliaceæ in Cretaceous rocks. It must, however, be admitted that the union of the stamens into a tube, so characteristic of the Bennettitean flower, is quite unlike anything we find in Magnoliaceæ or their near allies. This character has suggested to Dr. Wieland that the Mesozoic Cycadophytes may also have had an affinity with Dicotyledonous families much higher up in the scale, the *Convolvulus* family, for example. On this view a number of lines of evolution would have started in various directions from ancestors allied to the Bennetiteæ. At present all this is very speculative; at any rate, it is clear that the flowers of the Mesozoic Cycadophytes were highly organised structures; if it was from ancestors such as these that the higher Flowering Plants were descended, the evolution of the flower must have been often a process of reduction to greater simplicity, and the simplest flowers of Angiosperms would seem to be the furthest from the primitive type.

If, as seems to be now the most likely view, the Dicotyledons sprang from the great class of Cycadophytes which overspread the world in Mesozoic times, the question remains, what was the origin of the Monocotyledons? There is nothing to connect them directly with the Cycadophyta or any other Gymnosperms, and we can only suppose that they branched off from the Dicotyledons soon after the latter started as a line of their own. Much ingenious work has been done on this question, and it has been supposed that the peculiar characters of Monocotyledons (their general loss of secondary growth, for example) may have been due either to their adopting a largely underground mode of life (bulbs and corms), or to their having taken to the water at an early stage of their evolution. All this, however, is pure speculation; we really know nothing at all at present of the evolution of Monocotyledons; the clearly attested presence of so typical a family as the Palms well back in Cretaceous rocks is as yet quite unexplained. This is one among many indications that the ancestry of the Angiosperms as a whole must go back a good deal further than we have yet traced it, and that the Cycadophyta from which they appear to have been derived, must have belonged to the earlier Mesozoic Floras.

As we saw in the last chapter, the rise and progress of the Angiosperms was probably due, above everything else, to their adaptation to the contemporary insect life. It is an interesting question whether the relation between flowers and insects had already been established among the Mesozoic Cycads. The flowers of *Bennettites* must have been striking objects when fully expanded, rivalling, in size and modelling, the largest flowers of the present day. We can know nothing of their colours, but the bright tints which we often find in the fructifications of living Cycads suggest that similar attractions may not have been wanting in the flowers of their predecessors.

Even the cones of some members of the recent family are frequented by insects, which may probably serve as pollen-carriers. It is thus quite likely that the far more elaborate floral apparatus of the Mesozoic Cycadophytes may have already been adapted to insect-fertilisation.

However that may have been, the striking discoveries of Dr. Wieland have proved that the Cycadophyta of Secondary times possessed true flowers; in this and other respects they show a marked affinity to the Angiosperms which eventually displaced them. They have thus proved to fully deserve the name of Proangiosperms, which Saporta, by a brilliant inspiration, gave to *Williamsonia* and *Bennettites* at a time when their structure was very imperfectly known. We are still far from having solved the problem of the evolution of Angiosperms, but we have, I believe, the clue in our hands, and this clue leads us back to the highly organised Mesozoic Cycads as the great stock from which the modern Flowering Plants were an offshoot, destined soon to supplant the parent stem.

CHAPTER IV

THE EVOLUTION OF THE SEED-PLANTS

THE Cycadophytes of the Secondary period, as we saw in the last chapter, were very highly organised plants, their dominant family, the Bennettiteæ, far surpassing the modern Cycadaceæ and rivalling the Angiosperms themselves. This superiority of the old to the new might seem surprising if we took the obvious view that Evolution is a regular progression from the simple to the complex. This, however, is far from being the case; Evolution is to be compared to the successive waves of a flowing tide, rather than to the steady rise of a calm river. In Mesozoic times the Cycadophytes formed the wave that rose highest, but soon that wave receded, and another followed and rose higher still.

The Cycadophyta, in fact, were then the dominant class; there was nothing above them; they were the best thing in the way of Flowering Plants that their age had produced; from their triumphant success, overspreading all parts of the world, we cannot doubt that they were perfectly suited to the conditions then prevailing. Later on, the conditions became more complex, a change in which the increasing multiplicity of insect life was no doubt a chief factor. The Angiosperms, arising probably as an offshoot from the Cycadophyte stock, thus found their opportunity and rapidly fitted themselves into the numberless new places that the changing world presented to them. The higher members of the old family were no longer able to compete with their more enterprising younger line; the ancient race decayed, and, for all we know, became extinct; the less advanced Cycads did not come into such direct competition with the ascendant race, and have survived, in moderate numbers, as the recent Cycadaceæ.

A Mesozoic genus of Cycads, *Nilssonia*, of uncertain family, is known to have come down to late Tertiary (Miocene) times, in the island of Saghalien; Professor Nathorst suggests that *Nilssonia* may possibly still survive in China.

We will now pursue our inquiry a stage further back, and endeavour to trace the origin of the Cycadophyta themselves.

Both the great branches of the Cycad stock still show clear marks of affinity with Ferns, a fact which has long been recognised by botanists. As regards the recent Cycadaceæ, the mode of fertilisation (by multiciliate sperms), the arrangement of the pollen-sacs on the staminate scale, the circinate coiling of the young leaves in some cases and their venation in others, the leaf-like, stem-borne sporophylls of the female *Cycas*, are all characters which recall the Ferns. The Bennettiteæ, advanced as they were in many ways, yet in some respects approached the Ferns even more nearly than the Cycadaceæ, notably in their extraordinary compound stamens, so closely resembling the fertile fronds of a *Marattia* (see p. 147), and in the presence of a dense covering of flat scaly hairs, like the chaffy coat which clothes the young fronds of our present-day Ferns. It is probable that, as our knowledge extends, the resemblance will turn out to go further, for there are a number of Mesozoic fronds as to which it is still quite uncertain whether they belonged to Ferns or Cycads.

Our object is now to trace the Cycadophyta further back, and to see whether the Palæozoic rocks reveal anything of their evolution. If we can succeed in this quest, we shall clearly have made a long step towards finding the origin of the Seed-plants generally, for the Cycadophyta were not only of immense importance in the middle period of geological history, but, as we have seen, were the probable source of the now dominant race of Flowering Plants. The subject we have now to consider is thus of even wider scope than the origin of the Angiosperms, for it is the evolution of the whole body of Seed-bearing Plants which is in question.

We must now turn to the Flora of the Palæozoic period, by which, for our present purpose, we mean the Permian, Carboniferous and Devonian strata (see Table, p. 22), for we know little as yet of the plants of the still older rocks.

Just as the Mesozoic period has been called "the Age of Gymnosperms" so has the Palæozoic long been known as "the Age of Cryptogams." Certainly it is true that Spore-plants then formed a vastly more important element in the Flora than at any subsequent time; it is even probable that they were the main constituent of the great swampy forests which gave rise to the coal. Recent work, however, has shorn the Palæozoic world of Cryptogams of some of its glories.

The Flora of the Carboniferous epoch (by far the best known Palæozoic Flora) was, on a superficial view, made up of five great groups of the higher plants:

The Cordaiteæ, Gymnospermous trees;
The Horsetail family, which were also trees in those days;
The Sphenophylls, a group now extinct, distantly related to the preceding (see Chapter VII);

The Club-mosses or Lycopods, then the great arborescent family; and
The Fern-like plants.

The last mentioned were in number of species much the most important of all, amounting to about half the whole. It used to be assumed that this great mass of plants with the fronds of Ferns were really what they appeared to be—true Cryptogamic Ferns. But the researches of the last few years have confirmed a suspicion already felt by one or two clear-sighted investigators, that a large number of them were only Ferns in appearance and were really of a much higher rank in the Vegetable Kingdom. Only in the case of a mere fraction of these Fern-like plants is there any satisfactory evidence that they were true Ferns, reproduced by spores, like those of the present day. As regards the majority, the nature of the reproductive organs has only been in any degree revealed during the last seven or eight years; all the evidence goes to show that the great bulk of the apparent Ferns of Palæozoic age were really reproduced by seeds. In certain cases the seeds have been found in actual connection with the fronds; in other cases identity of structure enables us to recognize them; in a great number of instances, including most of the chief genera of the Fern-like plants, the association between particular fronds and particular seeds is constant. In fact, all the evidence we have (though direct proof is only afforded in a few cases) tends to prove that most of the Palæozoic plants which used to be classed as Ferns were in reality seed-bearing plants and not Cryptogams at all. Just as the Cycads of the Mesozoic have turned out, for the most part, to be much more advanced than recent Cycads, and very near the true Flowering Plants, so have the supposed Ferns of the Palæozoic proved for the most part to be far superior in organisation to the true Ferns and to have a real kinship with the Seed-plants.

What then were the affinities of these "seed-bearing Ferns" which formed so great a part of the Palæozoic vegetation? What was their relation to the Mesozoic Seed-plants on the one hand, and to the Ferns, which they so closely resembled, on the other? The discovery of a new and extensive class of Palæozoic seed-bearing plants may be expected to throw fresh light on the evolution of the higher groups.

We will begin by describing one particular type of seed-bearing Fern, and will choose *Lyginodendron oldhamium* (fig. 12), the first of these plants in which the seed was identified, and now the most completely known, as it happens, of all fossil plants. The piecing together of all the various, more or less fragmentary remains which belonged to *Lyginodendron*, so as to gain a connected view of the whole plant, has been a slow process; the foliage was described as long ago as 1829; the male organs, the last part to be recognised, not till 1905. We will not, however, follow historically the stages of reconstruction, but will briefly give the results as they now stand.

Lyginodendron oldhamium, a plant of the Lower and Middle Coal Measures, has been described as a little Tree-fern in habit; the stem was long and slender, seldom reaching two inches in diameter. It is possible that it may not have been upright, and that it supported itself with the help of stronger neighbours; the presence of spines all over the stem and leaves, like those on the climbing fronds of the West Indian Bramble Fern (*Davallia aculeata*), has suggested this idea; on the other hand there are small upright Tree-ferns, such as a Fijian species of *Todea*, with which the fossil plant might also be compared. In many specimens the stem was freely branched, but it is doubtful whether this was always the case; indeed it is quite possible that several distinct species may be grouped under the collective name *Lyginodendron oldhamium*.

The foliage consisted of very large and beau-

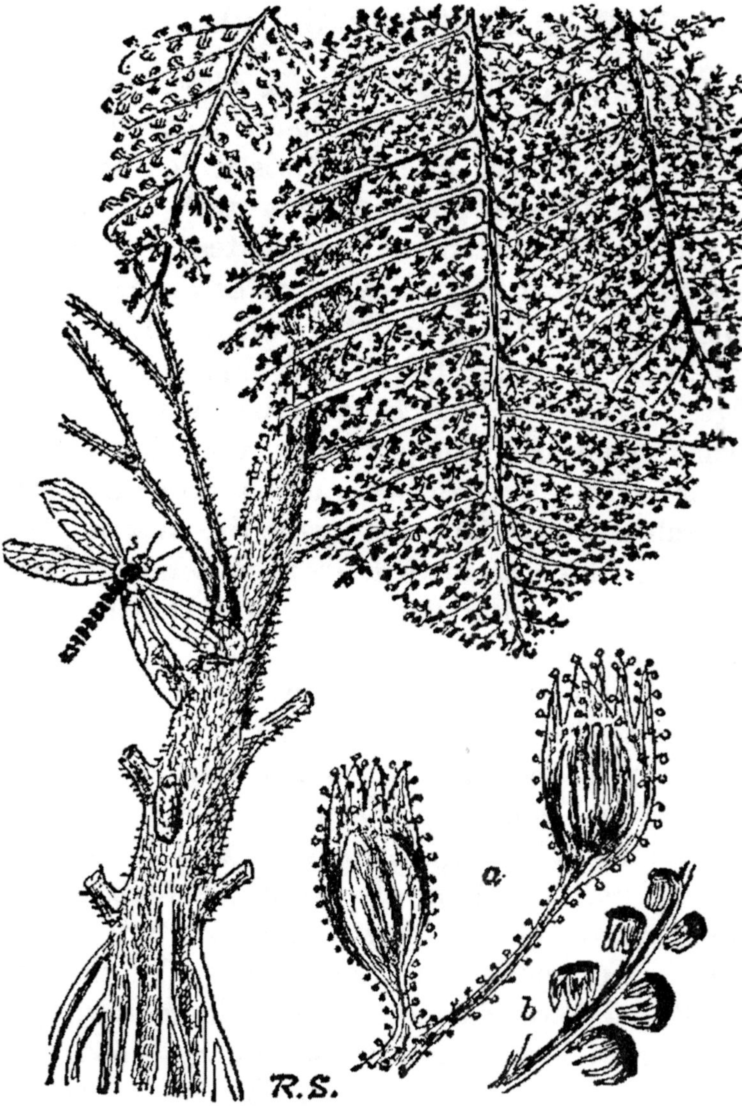

Fig. 12.—*Lyginodendron oldhamium*. Restoration, showing stem and roots, with the foliage; to the left above is a male frond. Observe the contemporary Dragon-fly. (*a*) Seeds in cupules, studded with glands, and borne on a branched stalk; (*b*) part of male frond showing discs and pollen-sacs. (*a*) and (*b*) slightly magnified. From drawings by Mrs. D. H. Scott.

tiful, much divided fronds, like those of some finely cut *Asplenium* or *Davallia*, but with the main stalk forked, a characteristic feature (see fig. 12). These great fronds were arranged spirally on the stem. Their external features were well known for many years before the stem which bore them was discovered. The little leaflets were rather thick, and curved in at the edges, characters which (together with the internal structure) suggest that the plant grew either in dry places or, as is much more probable, in a salt swamp. The stem gave rise to a number of roots, which seem to have been aerial in their upper part, growing downwards until they reached the soil and began to branch.

The anatomical structure of all these organs is quite well known, for fragments of all of them occur in abundance and beautifully preserved in the "coal-balls" or calcareous nodules of certain seams of coal in Lancashire and Yorkshire. We will not dwell on the anatomy here, for it would detain us too long. The structure of the stem when young was remarkably like that of the Royal Fern, *Osmunda;* with increasing age, however, a broad zone of secondary wood and bast was added by a cambium, the cells of which, delicate as they were, are often quite well preserved. In the older condition there thus came to be a considerable resemblance to the structure of a Cycad stem; in some details, especially the structure of the vascular strands passing out to the leaves, the agreement is remarkably exact. The leaf-stalk, however, only received a single strand, or at most a pair, which branched over and over again to supply all the complex subdivisions of the frond. The general structure of the frond internally was like that found in many of the Ferns. Besides the spines already mentioned, both stems and fronds were studded with spherical glands, usually stalked; we do not know whether they secreted honey or some other substance, but they have proved important as a means of identification.

Up to the year 1903 nothing certain was known of the reproduction of *Lyginodendron*, and some botanists thought that it was probably a true Fern in this respect, though so peculiar in structure. In that year, however, Professor F. W. Oliver was able to show that a previously unassigned seed, then known by a name of its own, belonged in reality to *Lyginodendron*, a result quite unexpected before. The seed, a small barrel-shaped body, about a quarter of an inch long, is enclosed in an envelope or cupule, like a hazel-nut in its husk (see fig. 12, *a*). This cupule bears glands, exactly like those on the frond and stem of *Lyginodendron*, with which it occurs in association; it was by the help of the glands that the seed was first identified; later on it was found that the stalk bearing the seed

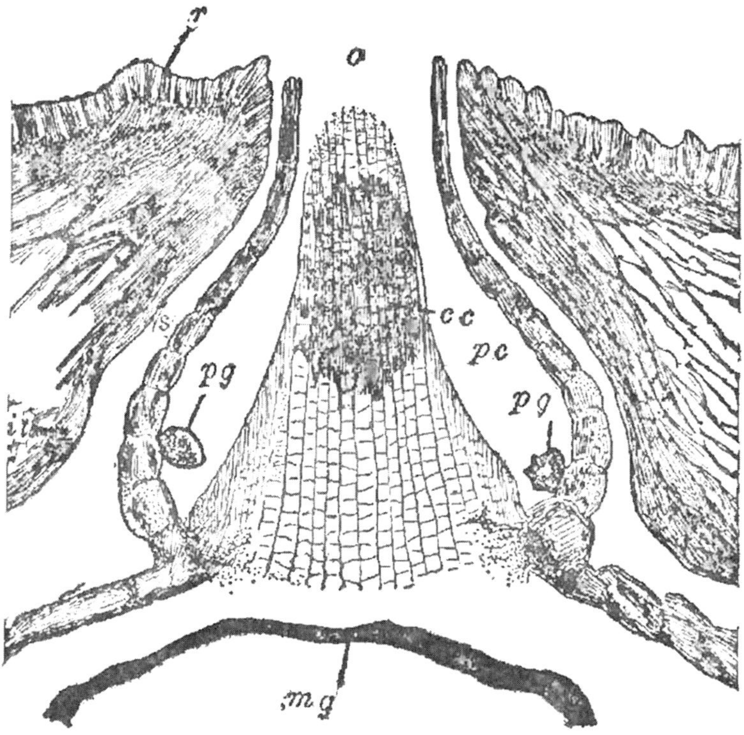

Fig. 13.—*Lyginodendron*. Longitudinal section of upper part of seed; (*r*) integument; (*s*) space between integument and pollen-chamber; (*o*) opening of pollen-chamber; (*cc*) central column; (*pc*) cavity of pollen-chamber; (*pg*) pollen in pollen-chamber; (*mg*) wall of megaspore. After Oliver. × 60.

had just the same structure as the stalks of the leaflets on the *Lyginodendron* frond.

The seeds are excellently preserved, so that it has been possible to make out most of the details of their structure. The seed itself, protected by the cupule or husk, consists essentially of a central body or nucellus, enclosed in a seed-coat; these two parts are closely united together except at the top. It will be remembered that in modern Cycads and in the Maidenhair-tree there is a hollow chamber in the apex of the nucellus serving to catch the pollen-grains. The same arrangement is present in the seed of *Lyginodendron*, and pollen is still found in the pollen-chamber; the latter, however, is less simple than in living Cycads, for a column of tissue rises up in the middle of the chamber, leaving only a narrow space round it for the reception of the pollen (see fig. 13, *pc*). It is interesting to find that the mouth of the pollen-chamber projected a little through the micropyle, so that it received the pollen directly, instead of the grains having to traverse the micropyle first. The seed-coat was supplied with a number of vascular strands, and had a complicated system of water-reservoirs in its upper part; they may have provided the water necessary for the purpose of fertilisation, if, as is probable, these plants were like Cycads in producing swimming sperms.

We do not know exactly how the seeds were borne in *Lyginodendron oldhamium*, except that the seed-stalk was evidently part of a leaf, as is shown by its structure. In other species, however, similar seeds have been found on the naked branches of a frond-like organ, so it seems that the seeds were produced on a part of the frond which did not have any leaflets (fig. 12, *a*). There was thus a certain difference between the sterile and fertile parts of the leaf, such as we find in the Royal Fern (*Osmunda*) and many other Ferns at the present day.

The discovery of these highly organised seeds showed at once that *Lyginodendron* was far removed from the true Cryptogamic Ferns, and was indeed already a Seed-plant, allied to the Cycadophyta.

But still our knowledge of the plant was incomplete, for nothing certain was known of the male organs. The pollen-grains had been seen in the pollen-chamber of the seed, but where had they come from? Mr. Kidston solved this problem in 1905, when he found the pollen-bearing organs in connection with the foliage of *Lyginodendron*. The polliniferous part of the frond bears, instead of the ordinary leaflets, little oval discs, two or three millimetres in length (see fig. 12, *b*); on the under side of each disc there are six or seven two-chambered pollen-sacs of a spindle shape. This type of fructification was known long before its connection with *Lyginodendron* was established, and as it occurs in a considerable variety of Fern-like plants of the period, we may infer that the family Lyginodendreæ was an extensive one. There is nothing in the pollen-bearing organs themselves to suggest that they belonged to a seed-bearing plant; in fact, until their connection with *Lyginodendron* was discovered, they were confidently referred to Ferns of the Marattiaceous group (p. 147).

We see then that *Lyginodendron* (which we may certainly take as the type of a great family of "seed-bearing Ferns") shows a very curious combination of characters. In the structure of the young stem and of the frond it was like a Fern; in the structure of the older stem and in other details it was like a Cycad. The seeds are quite highly developed and clearly rank with those of Cycads, but the pollen-bearing organs resemble the fertile fronds of a true Fern. The seed is, of course, the decisive character; *Lyginodendron* must be classed among the Seed-plants or Spermophyta, but it was more primitive than any Seed-plant of later periods, for both seeds and pollen-sacs were borne, as it appears, on little-altered ramifications of the ordinary vegetative fronds. Thus the *Lyginodendron* type goes even beyond the female plant of *Cycas* in the want of any special reproductive leaves (sporophylls) distinguished from the ordinary foliage. For this primitive group of Fern-like Seed-plants, of which *Lyginodendron* is an example, the name Pteridospermeæ (Seed-ferns) has been adopted.

One or two other examples of Pteridosperms may be given. The first case in which a seed was found in actual connection with a "Fern frond" was that of *Neuropteris heterophylla*, in which, in 1904, Mr. Kidston discovered large seeds, of the size of a hazel-nut, attached to a stalk which still bore the characteristic leaflets of the plant. The fronds of this "Fern" were of enormous size, comparable to those of the largest Tree-ferns, and in form somewhat resembled a gigantic *Osmunda*.

The family to which *Neuropteris* belonged is pretty well known, and a great variety of seeds have been referred to these plants, though the evidence is nowhere else so direct as in Mr. Kidston's species. They were mostly plants of large dimensions and rather like Tree-ferns in appearance, their trunks sometimes reaching a thickness of about two feet. The structure of the stem was peculiar, for instead of there being a single column of wood, as in all ordinary trees, there were here several such cylinders in the stem, each surrounded by its own zone of bast and growing in thickness by its own cambium. The many-cylindered structure is common in the stems of Ferns, as can be seen by cutting across the creeping stem of a Bracken-fern, but the combination of this character with indefinite growth in the thickness of each cylinder is peculiar to the fossil family and seems to us a cumbrous arrangement. The great leaf-stalks, with their numerous vascular strands and strong mechanical construction, resembled those of a recent Cycad. The plants were thus of great complexity in their vegetative structure.

As regards the reproductive arrangements, a number of seeds, which there is good reason for referring to this family, are well known. Generally speaking, we may say that the seeds of Neuropterideæ were very much like those of modern Cycads in dimensions and structure. They had a thick, fleshy outer coat, and a "stone" inside it, within which was the central body or nucellus, containing the embryo-sac. The nucellus had a well-marked pollen-chamber at the apex, much like that of a Cycad. The seed was well supplied with vascular strands, which formed a double system, one set running through the outer flesh, while another set traversed the nucellus. There are some differences of detail, but on the whole these seeds are the nearest among fossils to those of the recent Cycadaceæ.

A very interesting point in connection with seeds of this group is the wonderfully perfect preservation of the pollen-grains found in the pollen-chamber. It can be seen quite clearly that each pollen-grain contained a number of cells, with delicate walls separating them, which broke down to let out the contents. There can scarcely be a doubt that the bodies emitted from these cells were active spermatozoids; in fact, from the structure of some of these fossil pollen-grains, Renault deduced the probable presence of spermatozoids long before they were discovered in living Cycads or in *Ginkgo*. The many-celled structure of the pollen-grain extends to various other Palæozoic families. There is no proof, in any case, that a pollen-tube was formed, and it is quite likely that this structure had not yet been evolved, the spermatozoids relying more on their own powers of movement in the Palæozoic plants than is the case in the later Cycadophyta.

Besides the Lyginodendreæ and the Neuropterideæ there were a great number of other Fern-like plants of Palæozoic age which appear to have been reproduced by seeds. In two genera (*Pecopteris* and *Aneimites*) the seeds have been found in connection with the fronds, and are remarkable for having a flattened, instead of a rounded form; the importance of this point is that it shows a very near approach to the seeds of the Cordaiteæ, a Palæozoic family which had much in common with the Coniferæ.

Our knowledge of the Pteridosperms is still very imperfect, but nevertheless the discovery of this group has completely transformed our old ideas of the Palæozoic Flora. Most of the Ferns, so-called, were not Ferns at all, but seed-bearing plants, and hence the Palæozoic period can no longer be called the "Age of Cryptogams," in the sense that Cryptogams were then really dominant. The Seed-plants already held their own very well, and probably may have about equalled the Spore-plants in number of species.

What conclusion are we to draw from the presence, in Palæozoic times, of the vast group of Fern-like but seed-bearing plants which we call the Pteridosperms? It is not for nothing that they have been called Ferns for so long. The resemblance to Ferns is not only external — the two groups have many points of structure likewise in common, and, as regards the pollen-bearing organs, it is still almost impossible to distinguish between the pollen-sacs of a Pteridosperm and the sporangia of a true Fern. Both seeds and male organs appear to have been borne on almost unaltered foliage-leaves, an altogether Fern-like arrangement.

On the other hand, the characters allying the Pteridosperms to Cycadophyta are no less clear. There are many points of close agreement in the anatomy, so close as to have been quite enough to establish relationship before anything was known of the organs of reproduction. An even more important point is that the seeds have proved to be altogether of a Cycadean type, as shown by the most careful and detailed investigations, carried out by the earlier fossil botanists before the relation of the seeds to the so-called Ferns was known, and by Prof. F. W. Oliver and his colleagues since that discovery.

We are thus confronted with a great class of Palæozoic plants which combined, in the most striking way, the characters of Ferns and Cycads. We know that in the succeeding age Cycads, of varied and highly developed types, overspread the whole world. It seems clear that in the Pteridosperms we have traced them to their source or at least have carried the history of their descent a long stage further back. So far as any evolutionary conclusions from the still imperfect palæontological record are justified, we are warranted in the belief that the Pteridosperms represent the Palæozoic stage of evolution of the Cycadophyta and indicate, much more evidently than the Cycads themselves, that the whole of this great race of plants sprang ultimately from a common stock with the Ferns.

The study of evolution in the light of the Fossil Record thus shows us, what we could never have learnt from the living Flora alone, the extreme importance of the Cycadean stock. Represented now by a small and isolated family, the Cycadophyta were in Mesozoic times the leading class of plants, abounding, as their remains prove, in every part of the world, and attaining to the highest development which any plants had reached at that period.

When we go back to Palæozoic times, we find the Cycad line represented by a vast mass of forms, probably no less extensive than the Cycadophyta of the Mesozoic, very highly developed on their own lines, but more primitive, in so far as they were more Fern-like, than the Cycadophytes themselves. This race, on present evidence, is as ancient as any class of land-plants with which we are acquainted.

If, as we have found reason to believe, the Cycadophyta, in one or other of their branches, gave rise to the true Flowering Plants, it appears that the same great race has played a leading part throughout geological history, from the Devonian onwards, first as the Pteridosperms, then as the Cycadophyta, and finally as the Angiosperms which still prevail.

Probably some Pteridosperms lingered on into the Mesozoic period, but they were mostly replaced by Cycadophyta; in like manner some few Cycads have persisted, through the Tertiary, down to our own day, but as a dominant class they are replaced by the higher Flowering Plants. Each prevalent class in turn has given rise to a younger and more progressive branch, which has succeeded to the inheritance of the old stock.

Our attempts to picture to ourselves the march of evolution are necessarily very imperfect — we can only trace the successive scenes in the broadest outline.

In the history of the evolution of the Seed-plants there are two evident gaps. (1) We do not know how the transition from the Pteridosperms to the Cycadophytes was effected. The gap between a Pteridosperm and such a Cycad as the female plant of *Cycas*, is not, it is true, a very wide one, and we know that the *Cycas* type was of great antiquity, going back to the earlier Mesozoic. But though we may well compare the seed-bearing frond of a Pteridosperm with the leaf-like carpel of *Cycas*, while the seeds themselves offer no difficulties, yet, when we come to the male plant, there are already considerable differences, for we know of nothing in Pteridosperms, with their unspecialised, stem-borne sporophylls, at all like the Cycadaceous cone. Still less have we anything to show how the complex flower of the Bennettiteæ arose among the flowerless Pteridosperms. The stamens, indeed, with their frond-like form and numerous compound pollen-sacs are altogether Pteridospermous, not to say Fern-like, but the juxtaposition of the organs to form a flower, and the complex gynæceum with its reduced carpels are new characters, the origin of which still remains to be cleared up.

We happen to have but few plant-fossils from the Upper Permian or Lower Triassic rocks, and these blank pages of the geological history may account for the gap in our knowledge. Cycadophyta had begun to be evolved before the close of the Palæozoic period, but we know little of their fructification, and there is nothing to indicate any approach to the Bennettitean type. We may, however, mention here an extraordinary genus, *Dolerophyllum*, from the Upper Coal Measures and Permian of France and Russia, as an example of a quite peculiar early type of Cycadophyte.

The leaves were simple and round in outline, sometimes eight inches across; huge buds have been found, containing the young, unexpanded leaves. The structure of the vascular strands agrees exactly with that in the leaves of Cycads. Associated with the foliage-leaves, smaller discs have been found which have proved to be the stamens. These discs are very thick and contain a number of cylindrical chambers, placed vertically to the surface and filled with pollen-grains of large size and complex structure. Each grain was multicellular and opened by a lid.

Similar pollen-grains have been detected in the pollen-chamber of a seed of the same general type as those which are referred to the Neuropterideæ among Pteridosperms. The identification of the seed with *Dolerophyllum* is not, however, certain. The simple leaves and the strange stamens, with pollen-sacs completely buried in their tissue, separate this genus from all known Cycadophytes or Pteridosperms; the structure, however, both of the leaves and of the pollen-grains establishes a connection with one or the other group. While not forming a link in any known evolutionary chain, *Dolerophyllum* shows that at the close of the Palæozoic age the evolution of the Cycadophytes was going on in more than one direction. It may be added that in the same Permo-carboniferous rocks, stems have been found which are in some respects intermediate between those of *Lyginodendron* and Cycads, though not on the direct line leading to any known types.

The gap between Pteridosperms and Cycads is practically certain to be filled up by future research; probably the critical fossils may be already known, for there are various fronds, both late Palæozoic and Mesozoic, which may well have belonged to transitional forms; at present, however, the data are insufficient for tracing the connection in detail, though the affinity of the two great groups is clear.

(2) The other break in the evolutionary history of Seed-plants is more serious, and the circumstances less hopeful. This is the gap *below* the Pteridosperms. How were these plants derived from Ferns, and from what sort of Ferns did they arise? These are questions which may never be fully solved, for Pteridosperms appear to go back as far as any land-plants of which we have knowledge. Here again the affinities are clear; the habit, anatomical structure, and organisation of the pollen-bearing organs establish the closest relations with Ferns. We know of Pteridosperms which have altogether the anatomical structure of a Fern, and their pollen-sacs are often not to be distinguished from Fern-sporangia. It is when we come to the seeds that our real difficulties begin. Those of the Pteridosperms are very highly organised bodies and show very little trace of any relation to the spore-sacs of a Cryptogam. We must here pause to recall to our minds what are the essential differences between reproduction by seeds and reproduction by spores.

We may take an ordinary Fern as the type of the latter method in its simpler form. The spores, produced in spore-sacs (sporangia) on the frond, are all of one kind; each is a single cell, of rather small size. On germination, the spore gives rise to an independent green plantlet, the prothallus, on which the sexual organs, very definite and rather complex bodies, are borne. The male organ (antheridium) is a round, many-celled sac, producing a number of sperms in its central tissue; the female organ (archegonium) is likewise many-celled, and is a flask-shaped structure, with the egg-cell at the bottom, approached by a passage through the neck of the flask (*cf.* fig. 21, B and C, p. 185). Fertilisation requires the presence of water, for the actively moving male cells, the spermatozoids, to swim in. If this condition is fulfilled, fertilisation is fairly well insured, for the archegonia secrete a substance (malic acid) which is atttactive to the spermatozoids and guides them to their goal. The egg, when fertilised, develops into an embryo, which in time grows up into the spore-bearing Fern plant.

In the more advanced Spore-plants, *Selaginella* (so common in greenhouses), *Isoëtes* (a related genus, mostly aquatic), and the Waterferns (of which the Pillwort, *Pilularia*, is our native example) the sexes are distinguished at an earlier stage, for the spores themselves are of two kinds; there are numerous male spores (microspores) of very small size, and comparatively few female spores (megaspores) of relatively very large size. In the group of the Waterferns, only one large spore is produced in each spore-sac; in *Azolla* the spore-sac is enclosed in a special coat of its own and closely simulates a true seed or rather ovule.

In the microspores, the prothallus is scarcely developed; the spore has practically nothing else to do than to form an antheridium and emit the spermatozoids. On the female side, provision has to be made for the nutrition of the embryo, and here there is still a comparatively bulky prothallus, though it tends to lose the character of an independent plant, and to become a mere bearer of archegonia and storehouse of food-materials; it never frees itself from the megaspore in which it is formed (see fig. 18, p. 164). There are some obvious advantages in the differentiation of two kinds of spores. The male spores are kept small for easy dispersal, and can be provided in correspondingly large numbers; the prothallus, practically undeveloped in the small spores, is treated on economical principles and only produced where it is wanted, *i. e.* in the megaspore, in connection with the egg-cell from which the embryo is formed,

Since the time of Hofmeister, who laid the foundation of evolutionary botany, it has always been assumed that the condition with like spores (homospory) preceded that with unlike spores (heterospory). It is difficult to see how the order could be reversed, though it must be admitted that, so far as fossil evidence shows, heterosporous plants go back just as far as the homosporous. That, however, may be because the record has lost all its earlier chapters.

We may regard the differentiation of male and female spores (microspores and megaspores) as the first step towards the evolution of reproduction by seeds. In the Palæozoic, all the Lycopods (Club-mosses) of which the reproduction is sufficiently known, have proved to be heterosporous; so also were a portion of the Calamites (Palæozoic Tree-horsetails); curiously enough no Ferns with two kinds of spores have yet been discovered in the Primary rocks.

The heterosporous arrangement, as we find it in Cryptogams, has difficulties of its own, which to some extent balance its advantages. In order that fertilisation may be accomplished it is necessary for the two kinds of spores to germinate together, as well as in the presence of an adequate water supply. This necessary association of the large and small spores, which often differ in mass in the proportion of 100,000: 1, if left to chance, is a very precarious matter, for bodies of such different weights, if carried by the wind, will be unlikely to come to rest together. Hence a vast number of microspores are necessary—a somewhat expensive arrangement.

One means of ensuring greater precision in

bringing together the male and female spores was to retain the latter within the sporangium, so that the place of meeting might be fixed beforehand, at a definite point on the parent plant. Various devices thus became practicable for catching the small snores and keeping them in the right position till they discharged their spermatozoids. This stage is represented by some living species of *Selaginella*, in which microspores and megaspores meet, the spermatozoids are emitted and fertilisation takes place, within the megasporangium (see fig. 18, p. 164); an embryo may even develop before the megaspore is shed.

The retention of the megapsore within its sporangium and the ultimate shedding of the whole as one organ is, generally speaking, one of the great characteristics of the seed-method of reproduction. The imprisoned megaspore forms its prothallus and pollination takes place on the parent plant; so do fertilisation and the development of an embryo in ordinary cases, but in Cycads, embryo-formation may scarcely begin till after the whole organ is shed. In Palæozoic seeds an embryo has never been found, so it is probable that the whole development of the embryo and perhaps even fertilisation itself (as distinguished from the preliminary process of pollination) was postponed to a late stage, immediately preceding germination.

Besides the retention of the megaspore, there is a third very general characteristic of the seed; the megasporangium is enclosed in one or more coats or envelopes; we already find this in the still cryptogamic Water-fern *Azolla*. We will not discuss the doubtful question of the nature of the seed-coats; the most obvious comparison is with the indusium of Ferns, a membranous envelope which in some cases may enclose only a single sporangium. At any rate, the coats are an essential part of the ovule or seed, which is thus not simply a megasporangium, but a megasporangium *plus* something else. In the young stage, or ovule, the opening (micropyle) left by the seed-coats receives the pollen in Gymnosperms, and forms a passage for the pollen-tube in Angiosperms.

We may now return to the seeds of Pteridosperms. They are true seeds, for they retain the megaspore permanently within the nucellus, which is the equivalent of the megasporangium, and the nucellus is enclosed in one or more coats. All the parts often have an elaborate structure, quite equal to that of the Cycadean seed, and there is little to show that these ancient seeds stand near the Cryptogamic sporangia. It is true that in the seed of *Lyginodendron* the neck of the pollen-chamber projected beyond the seed-coats, and no doubt caught its own pollen, without waiting for it to be brought in through the micropyle. This may be a primitive character, but it does not seem to have been general in the seeds of the Pteridosperms. The only important peculiarity of the Palæozoic seeds is the absence of an embryo, indicating that the nursing of the infant plant was a function which the seed did not take on till later times.

In all other respects the seeds of the Peteridosperms were true seeds; the stages of their evolution from the female sporangium, from which we suppose them to have been derived, are still to be discovered. Some light is thrown on the question by another line of descent, that of the Club-mosses or Lycopods, some of which produced a kind of seed, a subject to which we shall return in Chapter VI.

On the whole, the seed-bearing Ferns may be regarded as showing a relatively early stage in the evolution of Seed-plants, though the seed itself was already so far advanced. We must not, however, suppose that the Pteridosperms by any means represented the highest stage of evolution reached in the Palæozoic Flora.

A contemporary group, known to Science long before the Pteridosperms were recognised, attained a much higher development than they, rivalling the Conifers of later periods. This was the family of the Cordaiteæ; as regards some forms of the type-genus, *Cordaites*, our knowledge is very complete, thanks to the brilliant work of Grand' Eury and Renault, who, more than thirty years ago, succeeded in reconstructing these

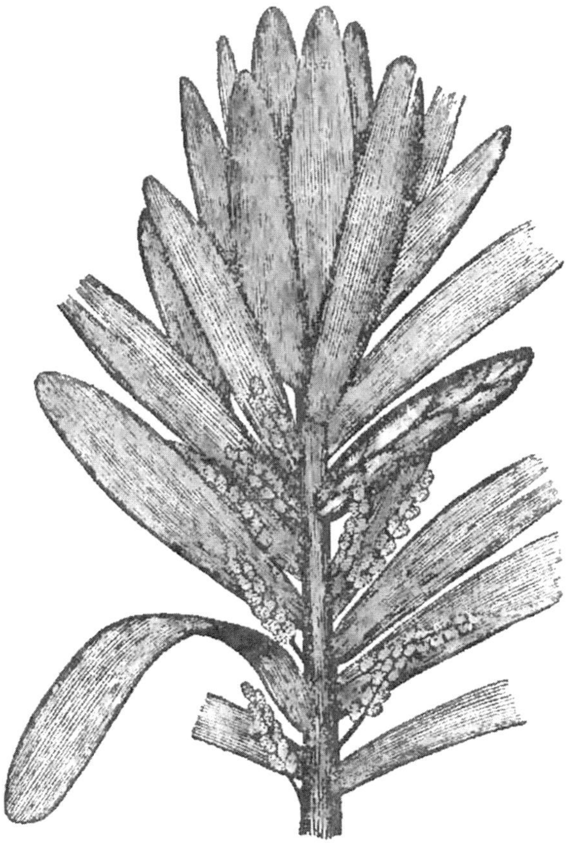

Fig. 14.—*Cordaites.* Branch (restored), showing leaves, bud and catkins. After Grand' Eury.

plants from their fragmentary remains. Our knowledge has not increased since that date so fast as might have been hoped.

The family was evidently an extensive one, and goes back to Devonian times. The stems, the stumps of which have been found rooted, as they grew, were lofty trunks, reaching a height of 100 feet, ending in a crown of branches bearing rather narrow leaves, sometimes a yard long (see fig. 14). In general appearance the trees most nearly resembled certain Conifers of the Southern Hemisphere, related to the Kauri Pine of New Zealand; some of those have leaves six or seven inches long by two inches broad—the largest in living Conifers, though far inferior to those of the ancient group. In structure the stems of *Cordaites* likewise resembled this southern family of Conifers, except that the pith was much larger and chambered like that of a Walnut-tree. On the other hand, the leaves had almost exactly the same structure as the leaflets of a Cycad such as *Zamia*, though on a much greater scale.

The reproductive organs, like the vegetative, were in many respects much more advanced than those of the Pteridosperms. The latter, as we have seen, though complex enough in their own way, resembled the Ferns in bearing both seeds and pollen-sacs on fonds or parts of the frond differing little from the ordinary foliage; this was far from being the case with the Cordaiteæ. Their fructifications were cones or catkins, quite as sharply differentiated from the ordinary branches as in a modern Coniferous tree (fig. 14).

In the male catkins, which were borne in clusters, the stamens were protected by bracts. Each stamen consisted of a long stalk bearing at the end from four to six long pollen-sacs, which stood erect. The best comparison perhaps is with the male catkins of the Maidenhair-tree. Here the stamens are likewise stalked, and usually bear two pollen-sacs, which, however, hang down instead of standing erect. Sometimes there are three or four pollen-sacs even in the recent plant, and the larger number is also found in Mesozoic Maidenhair-trees. The intermingling of bracts with the stamens is, however, peculiar to *Cordaites*.

The female fructifications differed little from the male in outward appearance. The ovules, few in number, were borne, among the bracts, each at the end of a short stalk. By singular good fortune, the ovules, as well as more mature seeds, have been found well preserved. There is a well-developed pollen-chamber, in which the pollen-grains are formed; in fact, *Cordaites* was the first fossil plant in which the many-celled structure of the pollen was discovered. The ovule and seed were altogether much like those of recent Cycadaceæ, and at the same time differed little in essentials from some of the seeds of Pteridosperms. The form of the seed is flattened, which is also the case in some members of the latter class. It is not possible at present to distinguish with certainty in all cases between the seeds of Pteridosperms and those of Cordaiteæ. This is an imnortant point, for in most respects these great trees bore little enough resemblance to the Seed-ferns, A fairly complete series of Palæozoic stems, however, is known, connecting the stem-structure of *Cordaites* with that of such a Pteridosperm as *Lyginodendron*, while the structure of the leaves, as already mentioned, is like that of a Cycad, though they were simple instead of compound in form. On the whole, it is evident that the Cordaitcæ were related to the Pteridosperm-Cycadophyte line of descent; the enormous antiquity of such highly organised Seed-plants is a most striking fact; their connection with the Seed-fern stock must lie very far back, in Lower Devonian times, if not earlier.

It thus appear that during Palæozoic ages the Seed-plants were represented by at least two great classes, both of which go back as far as our records of a Land-flora extend. On the one hand, there were the Seed-ferns, probably the more extensive group of the two; plants which, while retaining many of the characters of the Fern stock, had already evolved a high type of seed. On the other hand, there were the Cordaiteæ, on a level with the Seed-ferns as regards the seeds themselves, but in every other respect more advanced. As Saporta, the French palæobotanist, recognised when the Cordaiteæ were first made known, their reproductive organs exceeded in elaboration even those of the Coniferæ, but were on a different and essentially more primitive line, and one doomed to extinction.

It is an interesting question, what became of this great Palæozoic group of trees? Did they really become extinct, or have they left any survivors? In the form in which we know them they no doubt died out, but very gradually, lingering on far into the Mesozoic.

Their nearest living representative is undoubtedly the Maidenhair-tree, connected with the Cordaiteæ by many characters, especially the structure of both male and female fructifications, including the detailed organisation of the plum-like seeds. The Maidenhair-tree line, so well developed during Mesozoic times, probably did not spring directly from those Cordaiteæ which happen to be best known to us at present, but was no doubt derived from some other branch of the same great class.

A more important and difficult question is that of the relation of Cordaiteæ to the Coniferæ.

It is clear that Saporta in a general way was right; the two groups are distinct, as shown especially by the seeds of the former, which are of the Cycadean, not the Coniferous type. On the other hand, in habit and the structure of the wood the Cordaiteæ come nearer to the Araucarian Conifers (*i. e.* the family to which the Kauri Pine and the Puzzle Monkey belong) than to any other plants; the stamens of the latter, with their numerous and long pollen-sacs, recall in a certain degree those of the Palæozoic family. A detailed comparison between the two groups would take us too long; there are, however, so many points in common between them that there seems little reason to doubt that there was a real affinity, as was believed by the botanists to whom our present knowledge of the Cordaiteæ is due. The Coniferæ themselves appeared in late Palæozoic times, and their earliest representative, the Permian genus *Walchia,* was probably an Araucarain. The most tenable view appears to be that the Coniferæ sprang from the same class to which the Cordaiteæ belonged, though not from that particular family.

The Palæozoic period was once called the Age of Cryptogams. If we could see the luxuriant forests of the Coal period, we should probably fancy this description just; the vegetation must to a great extent have presented the aspect of that in the Tropical Fernhouse of a Botanical Garden, so dominant were forms like those of the Ferns and Fern-allies.

One class of plants, however, would have stood out from all the rest, anticipating the Flora of a later age. The tall Cordaitean trees had already long cast aside all traces of the Cryptogamic habit and revealed themselves in every character as advanced and typical Seed-plants.

Among the rest of the vegetation, plants with the appearance of Ferns were the most numerous in species, and formed, probably, the vigorous undergrowth of the forest, sometimes themselves reaching the stature of trees. The majority of the "Ferns," though there was little in their outward aspect to betray their real nature, would have proved to bear on their fronds, not the spore-sacs of a Fern, but true seeds, as perfect as those of the Cordaiteæ themselves.

Of these two great Palæozoic seed-bearing races, the advanced Cordaiteæ, it appears, were the forerunners, if not the direct ancestors of the Coniferæ; the fern-like Pteridosperms had an even greater future before them, for they were destined first to give rise to the world-wide Cycadophyte Flora of the Mesozoic ages, and then, in still later times, to become the ancestors of the true Flowering Plants which have for so long taken the leading place in the vegetation of the Earth.

CHAPTER V

THE EVOLUTION OF THE HIGHER SPORE-PLANTS—THE FERNS

We have so far been considering the subject of the Evolution of the Seed-plants, the question which is of primary interest to us, because these are the highest plants and beyond comparison the most important in the Flora of the World as it now is. The conclusion we have reached is that the Seed-plants were derived, in very early times (perhaps Lower Devonian, perhaps earlier) from Spore-plants of the same main stock with the Ferns.

The Spore-plants, or Cryptogams, as they are traditionally called (the old Linnean name, implying that the sexual reproduction of these plants was *hidden*, which is no longer the case), are still with us, in large numbers. We cannot attempt to deal with them all, but it will be of interest to trace what we can of the evolution of those groups of Spore-plants of which alone the history is adequately known, namely, the higher families, which stand nearest to the Seed-plants. Their importance depends partly on this relation, partly on the fact that some of their races were incomparably better developed in early geological times than at present, so that their past history is well worth tracing.

We may shortly define the *higher* Spore-plants, with which alone we are concerned, as those in which there is a regular alternation of sexual and asexual individuals, the individual bearing the sexual organs being always relatively small and simple (the prothallus), while that bearing the spores is highly developed and constitutes the *plant* itself, as ordinarily known to us. Thus in a Fern, a *Lycopodium* or a Horsetail the plant is strictly asexual, bearing the spores, which in these cases are all of one kind. The prothallus or sexual generation, bearing the antheridia and archegonia (see above, p. 121), is always small and may never have been seen even by most people to whom the plant itself is familiar. In fact, in the case of *Lycopodium* and the Adder's Tongues, it is only within the last quarter of a century or less that the sexual individual has become known to science.

In certain of the higher Cryptogams already referred to (p. 122), namely, in the peculiar group of the Water-ferns, and in *Selaginella* and *Isoëtes* among the allies of *Lycopodium*, a certain sexual distinction is already manifest on the plant itself, for here the spores are of two kinds. In these cases the alternation of generations becomes somewhat obscured, for on the male side the prothallus is reduced to almost nothing, while the female prothallus, though fairly large, remains to a great extent enclosed within the megaspore. In the Seed-plants, as already explained, this reduction of the sexual individual goes still further, the female prothallus, together with the megaspore, remaining permanently enclosed in the ovule and seed. At the same time, the organs producing the two kinds of spores—the stamen with its pollen-sacs on the one hand, and the carpel with its ovules on the other—come to be strongly differentiated, so that the terms male and female are commonly applied to them.

To return to the higher Spore-plants, in which the characteristic alternation of generations is more evident, we find that in the living vegetation of the World they are represented by three great groups, the Ferns, the Club-mosses or Lycopods, and the Horsetails or Equisetums.

The Ferns, familiar to every one, are characterised by the great development of the leaf in comparison with the stem; the frond is the typical part of a Fern. The numerous spore-sacs are borne either on the ordinary fronds (usually on the back) or on special fronds or leaflets; a cone is never formed. Most Ferns have spores of only one kind; the small family of the Water-ferns alone bears spores of two kinds.

The Club-mosses, unlike the Ferns, have small and simple, but numerous leaves, borne, as a rule, on long and conspicuous forked stems, the habit being like that of a Moss on a large scale. The chief exception is *Isoëtes*, a genus mostly aquatic, in which the stem is much stunted, and the leaves, though simple, are comparatively long. The spore-sacs are invariably borne singly on each leaf, either on its upper surface, or in the angle between leaf and stem. Sometimes the leaves associated with the spore-sacs are ordinary leaves; more often they are special sporophylls, grouped together to form a definite *cone*. The class includes many plants with one kind of spore, and still more numerous species in which microspores and megaspores are differentiated. *Lycopodium* and *Phylloglossum* represent the former, *Selaginella* and *Isoëtes* the latter condition.

The Horsetails are widely different from both the other classes. The leaves are extremely reduced, appearing as mere teeth on the sheath formed by the united leaf-bases. The stems are well developed and jointed, the leaves and branches being in whorls. The cones are very sharply marked off; the spore-sacs are borne on the under side of shield-like scales; the spores are all of one kind.

We will begin with the Ferns, because they appear to come nearest, as we have seen in the last chapter, to the line of descent of the Seed-plants.

Our modern Ferns, indeed, show little sign of their high connections, and it is only from the study of fossil plants that the evidence has come. The Ferns which have come down to our own day are frankly Cryptogamic, and for the most part make no approach to higher groups. The exception is the family of the Water-ferns, in which, as we have seen, there is a very marked distinction between the two kinds of spores; in fact, the most extreme cases of heterospory among living Cryptogams are in this family. The interesting point is that in the megasporangium only a *single* spore comes to maturity, filling the whole sporangium, like the embryo-sac of an ovule. Where, as in the little floating water-weed *Azolla*, an envelope grows up round the sporangium, the resemblance to an actual ovule becomes surprisingly close. There is, however, no formation of a seed, and fertilisation takes place after the spores have been discharged into the water. Probably the Water-ferns have really nothing to do with the evolution of the Seed-plants; they appear to belong to a later branch altogether, which to a certain extent shows an interesting parallelism with the main line of descent.

Most of the families of our living Ferns can be traced well back into the Mesozoic period;

the most familiar Ferns, the Polypodiaceæ (*e. g.* Bracken, Male Fern, Hart's Tongue, etc.), have not been found in older rocks than the Jurassic. We will only say a few words on the history of two families which are of special interest—the Osmundaceæ, or Royal Ferns, and the Marattiaceæ, tracing them back, as far as possible, to their Palæozoic source.

The Royal Fern, a handsome plant, growing in bogs and wet woods and especially common in Ireland, is only too well known, for its popularity with collectors and gardeners threatens its extinction in many places. It has an extraordinarily wide distribution, extending from Europe to Japan, India, and South Africa, and from Canada to South Brazil.

In this plant the upper part of the frond is fertile, the leaflets being reduced to mere stalks, crowded with sporangia; the lower part bears the ordinary green leaflets. In other species of *Osmunda* and even in some forms of *O. regalis*, there are separate vegetative and fertile fronds, while in *Todea*, a Southern Hemisphere genus, there is no distinction at all, the sporangia appearing on the under side of unaltered leaflets.

The sporangia themselves are different from those of most modern Ferns, such as the Bracken, the Male Fern or the Maidenhair. In the latter a single row of thickened cells forms a ring round the sporangium and by its contraction breaks open the wall and scatters the spores. In Osmundaceæ, instead of the ring, there is a little patch of thickened cells on one side of the sporangium. This, as we shall see, marks an old type of sporangium, which in the Royal Ferns has come down to our own time. The stem, short, thick, and almost subterranean in our native species, but tall and erect in some Southern forms, has a structure unusual in Ferns, recalling that of a Dicotyledon. There is a ring of vascular strands surrounding a pith; the wood of the strands is separate, but the bast forms a continuous zone all round the outside of the ring. A single strand passes out into the petiole of each leaf. In an American species a little bast is found at certain places on the *inner* side of the wood, and it is an interesting fact that in a Cretaceous species, from Queen Charlotte Islands, British Columbia, the same peculiarity is found in a more extreme form; here, there is a regular band of bast to the inside as well as the outside of the wood. Thus the Osmundaceous stem seems to have reached its greatest complexity in Cretaceous times.

Fortunately there is a series of fossils, clearly belonging to this family, which enables us to trace back its history to the Palæozoic period, as has been done by Kidston and Gwynne-Vaughan. As we go back to the older forms the tendency is for the ring of separate strands to be replaced, first, by a continuous zone, and then by a solid cylinder of wood; as we should expect, the series is not perfectly regular, the different types sometimes overlapping. Thus normal *Osmunda*-like stems have been found, not only in the Tertiary, but as far back as Jurassic rocks. In a Jurassic stem, however, from New Zealand, the zone of wood is almost continuous, and in a species from South Africa, of about Wealden age, there is the further peculiarity that wood-vessels occur inside the ring, scattered among the cells of the pith. All these forms are referred to the genus *Osmundites*.

Some extremely interesting stems are recorded from the Upper Permian of Russia, showing what the Osmundaceous structure was like at the close of the Palæozoic period. It is rare to find fossil plants at this horizon, and in this case the result justifies our curiosity as to the Flora of these rocks. In one species, *Zalesskya gracilis*, the wood forms a perfectly continuous, very broad zone, with no sign of division into separate strands. A very curious point is that while the outer part of the zone has the ordinary wood-structure, the inner consists of shorter and wider vessels, such as we find, in recent plants, used for storing water, rather than for conducting it up the stem. In this form there was still a pith, but in another species of *Zalesskya* it is probable that the inner, large-celled wood extended to the centre, while in the magnificent Fern *Thamnopteris*, the wood certainly formed a solid cylinder with no pith at all, the middle being occupied by the short and wide wood-elements already mentioned. We have here got back to a structure of the vascular system very different from what we know in recent Osmundaceæ, but the other characters, especially the structure of the petiole, leave no doubt as to the relationship. The interesting thing is that these oldest Osmundaceous stems closely resemble some stems of the most ancient known group of Ferns, the Primofilices, characteristic of Palæozoic rocks.

Thus a modern family has been traced back till it almost becomes merged in an ancient group, which probably represents the common source of most of our living Ferns, as well as of the Osmundaceæ. The evidence for the history of this latter family does not rest only on anatomical structure, important as it is. In various Mesozoic rocks, down to the Rhætic and Triassic, Ferns have been found with the characteristic sporangia of Osmundaceæ, and it is a most interesting fact that sporangia of the same general type are common in petrified material of the Palæozoic period. We might call this the most ancient type of Fern-sporangium; it occurred not only in fossil Osmundaceæ but in their predecessors, the Primofilices, so that evidence from the reproductive as well as the vegetative organs connects the recent family with this ancient group, of which a slight sketch will now be given.

The effect of the discovery of the Pteridosperms has necessarily been to reduce greatly the number of true Palæozoic Ferns. There remain, however, two considerable groups of plants of that period, which may be accepted as Ferns. The simpler of the two groups we have called, following Mr. Arber, the Primofilices, because they appear to go further back in geological history than any others of their class. They are best known from their structural characters, for there are not many instances in which we have good evidence for the external habit of members of the group, though doubtless some of the fossil impressions of Fern-fronds really belonged to them.

Most modern Ferns have a number of distinct vascular cylinders in the stem; the Primofilices, like the Royal Ferns and one or two other recent groups, always had a single vascular cylinder; there was also one bundle only in the petiole of the frond. In some, the stem had a remarkably simple structure, the single vascular cylinder consisting merely of a solid column of uniform wood, surrounded by a layer of bast; in others, the central part of the wood was different from the outer zone, as in the Permian Osmundaceæ; and in others again, the centre was occupied by a mixture of wood and pith, as in the Wealden *Osmundites* mentioned above. In these more complex cases the woody column was apt to lose its simple cylindrical form and to become fluted and grooved, often to an extreme extent. Sometimes the branching of the stem was a mere bifurcation, as it is in the modern Bracken, but in other cases the branches were axillary, springing from the angle between stem and leaf, as happens in Filmy Ferns at the present day, though otherwise not a Fern-character. The petioles were in many cases thicker than the stem that bore them, and the single vascular strand of the petiole often had a complicated structure. A frequent peculiarity is that two rows of leaflets were given off on each side of the main leaf-stalk; this strange arrangement, however, may only have belonged to the fertile fronds, for it was commonly the case in this group that the sporangia were borne on special fertile fronds without leaflets; in this respect these early Ferns were already rather advanced, compared with most of their present-day relations. In other cases, however, the fructification was on the under side of ordinary leaves.

The sporangia, as already mentioned, were often much like those of the Osmundas, having a patch of thickened and enlarged cells on one side. In other cases greater elaboration was at-

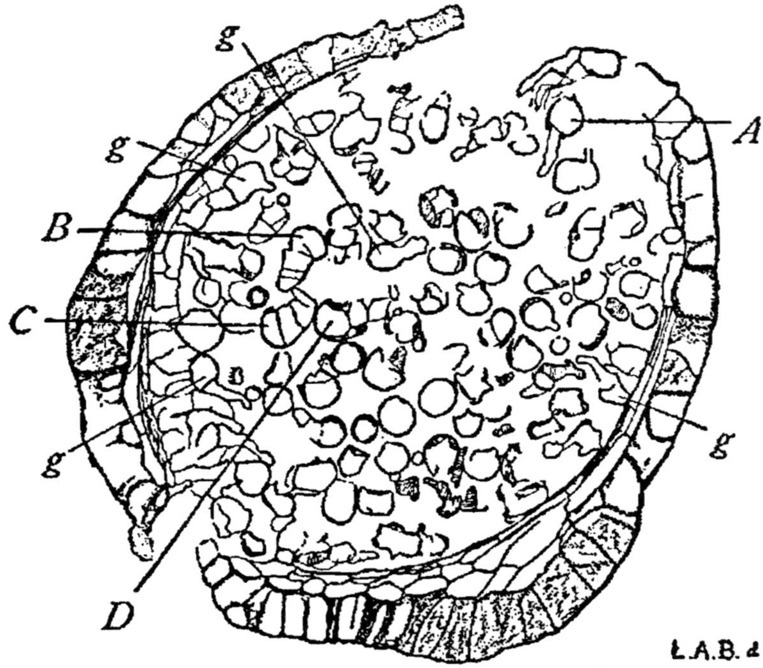

Fig. 15.—*Stauropteris.* Sporangium containing numerous spores, some of which, at the places lettered, are germinating. × about 70.

tained; the sporangia were of large size and the apparatus for discharging the spores was on a corresponding scale. Sometimes there was an approach to the ring-mechanism of most modern Ferns, but the ring was usually, if not always, more than one cell in width. In other species again there was no special mechanism of the kind at all.

The spores, so far as has been observed, were all of one kind, and correspond in dimensions to those of recent Ferns.

The same question might arise here as in the case of other Palæozoic plants resembling Ferns —were they true Ferns or not? It so happens that we have evidence of an unusual kind bearing on the point. In more than one case the spores still enclosed in the sporangia of Primofilices have been caught in the act of germination, and are so well preserved that the details of the process are clear (see fig. 15). The stages observed agree quite well with corresponding early stages in the development of recent Fern-prothalli. It is well known that in modern Ferns (including members of the Osmundaceæ) germination not uncommonly takes place inside the sporangium, if, owing to damp weather or other causes, the spores have failed to be discharged. The cases observed in the fossils may have been equally accidental; if so, it was a fortunate accident for botanists; the evidence of the germinating spores, in connection with all the other characters of the plants, can leave no reasonable doubt that the Primofilices were really an early race of true Ferns. As we have seen, they show direct continuity with the family Osmundaceæ; they also have points in common with several other families of recent Ferns, and it is probable that the Primofilices, taken in the widest sense, represent the Palæozoic stock from which most of the later groups were derived in Mesozoic times.

The Primofilices, however, by no means exhaust the Palæozoic Ferns; another and more

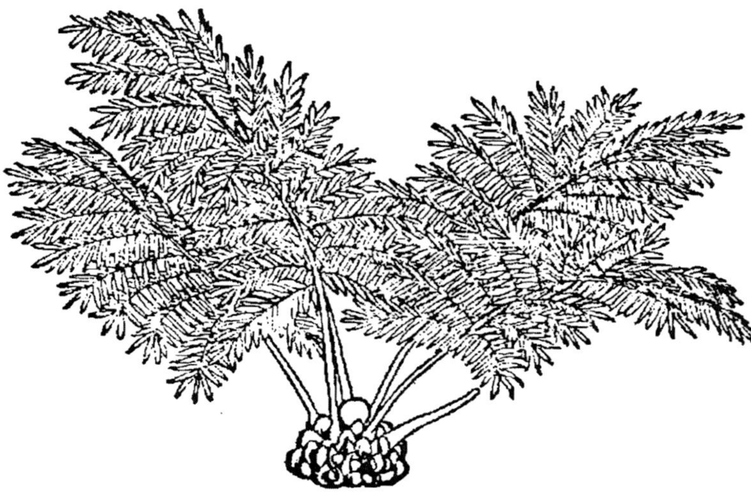

Fig. 16.—*Angiopteris evecta.* After Engler and Prantl.

complex family has still to be considered, and in this case also we will approach the fossils through a recent group. Unfortunately the family Marattiaceæ, with which we are here concerned, has no popularly known members, though some of them will have been noticed by those who have visited the Fern-houses at Kew. They are almost entirely tropical Ferns, a few spreading into temperate regions in New Zealand and Southern China. Many of them are fine, handsome Ferns, the fronds reaching a length of twelve feet in *Angiopteris*, the one most common in cultivation. The stems are usually short and massive; the fronds, commonly much divided, are smooth and almost free from the chaffy scales usual in other Ferns (fig. 16). A characteristic point is the presence of two fleshy outgrowths (stipules) at the base of the petiole.

The structure of the stem is very complex, the vascular system consisting of a number of cylinders of various forms, often ranged in two or more concentric circles, and fusing with one another to form an elaborate network. Each leaf receives a large number of vascular strands from the stem. The roots, which are themselves larger and more complex than those of other Ferns, arise near the apex of the stem and burrow down through the cortex till they become free and reach the soil. In their vegetative structure the Marattiaceæ are perhaps the most highly organised of all Ferns; their reproductive arrangements are also of a high type.

In most Ferns, it will be remembered, the sporangia, though commonly grouped together in definite clusters or "sori," are quite distinct one from another. In the majority of the Marattiaceæ a number of them are joined together to form a compound spore-sac, or "synangium." In *Angiopteris* and one other closely allied genus, this is not the case; the sporangia are very closely associated in two ranks to form each "sorus," but are not actually joined. Even here, however, the sporangia are different from those of most Ferns; they are larger and of more solid construction, with a wall several cells thick. The opening of the sporangium to set free the spores is effected by means of a patch of thickened cells at the apex, an arrangement bearing some resemblance to that in the Osmundas. In the other three genera, *Marattia*, *Danæa* and *Kaulfussia*, including the great majority of the species, the sporangia are united. In *Marattia* they form pod-shaped fruits or synangia, each containing a double row of compartments (of which there are about ten in each row), representing the sporangia (see fig. 17, A). In one species the whole compound body is stalked, and closely resembles a compound

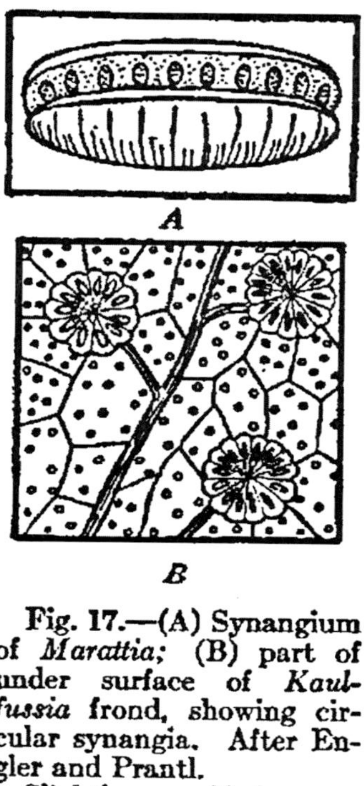

Fig. 17.—(A) Synangium of *Marattia;* (B) part of under surface of *Kaulfussia* frond, showing circular synangia. After Engler and Prantl.
Slightly magnified.

pollen-sac of *Bennettites*. When ripe, the two kidney-shaped halves of the fruit split apart, like opening a book, and then each of the numerous compartments in either half releases the spores by means of a slit on its inner surface (fig. 17, A). *Danæa* has much the same arrangement, except that the spore-fruit does not split open and the spores escape through pores instead of slits. This genus shows some slight difference between sterile and fertile fronds, which are not usually distinct in the family.

In *Kaulfussia*, which has fronds like the leaves of a Horse-chestnut, the synangia scattered on the under side are circular, each having the form of a neat rosette, with a cup-shaped hollow in the middle, into which the compartments (about a dozen in number) open by slits (see fig. 17, B).

These curious spore-fruits are no doubt an advanced form of fructification compared with the simple spore-sacs of the ordinary Ferns. Yet the family Marattiaceæ, so highly-organised both in vegetative and reproductive structure, represents a very ancient race. A certain number of Mesozoic Ferns have been referred to this family, and some of these, from rocks of Triassic and Liassic age, are so like the recent forms that they have been placed in the genera *Marattia* and *Danæa*. When we get back to Palæozoic times we meet with a surprising number of Ferns, which, both in their structure and fructification, show manifest kinship with Marattiaceæ—in fact this group has been regarded as the dominant family of Ferns in the later Palæozoic ages. The evidence is remarkably complete; both roots and leaves are found in connection with the stems, and the internal structure of the latter is admirably preserved in a great number of different species. On fronds which agree in structure with those of which the bases still adhere to the stems, the fructification has been discovered in numerous cases. Perhaps no group of fossil plants is better known in all its parts than this, though, since the discovery of the Pteridosperms, certain difficulties have arisen in the interpretation of the facts.

The stems differed from those of recent Marattiaceæ in stature; the Palæozoic forms were Tree-ferns, reaching, it is said, a height of 60 feet. The anatomical structure was, as a rule, extremely elaborate, even more so than in the recent family, but of the same type. The stems were enveloped in a dense mantle of roots, agreeing perfectly in structure with those of the living Marattiaceæ. The great fronds, however, had a simpler vascular system.

The fructifications are the feature of special interest. Synangia are found on the fertile fronds in considerable variety. A large proportion are of the circular, disc-like form now only represented in the genus *Kaulfussia*. In some of the fossil spore-fruits the sporangia were only connected at the base, or they may even have been nearly as free as in the recent *Angiopteris;* in others they were completely united, as in *Kaulfussia* itself. Sometimes the synangium was stalked, as in a part of the recent genus *Marattia*. In other cases there was a close agreement in form with the fructifications of *Danæa* and *Angiopteris*.

On the whole of the characters, it seems to be perfectly established that in Palæozoic, especially late Palæozoic times, there was a large group of plants, extending probably to all parts of the world, which had a clear affinity with the recent Marattiaceous Ferns.

The question that still remains is: were these Palæozoic forms true Ferns or were they Pteridosperms? The pollen-bearing fronds of *Lyginodendron* were at one time classed among Marattiaceous Ferns; another supposed member of the latter family has been found laden with seeds! Even in the highly-developed Bennettiteæ of Mesozoic age, the stamens might well be taken for the fertile fronds of some unknown member of the Marattiaceæ. There is, in fact, no means of telling for certain whether a given fructification represents the sporangia of a true Fern or the pollen-sacs of a Seed-fern. On the other hand, the anatomical features of the Palæozoic allies of Marattiaceæ are altogether those of Ferns, and have none of the peculiarities of Pteridosperm anatomy. On the whole, the probability is that a considerable proportion of the supposed Palæozoic Marattiaceæ were really Ferns of that group, though some have turned out to be seed-bearing plants. The difficulty in distinguishing between the two is in itself very interesting, and suggests irresistibly that there was a real affinity between Marattiaceæ and the early seed-bearing plants. Long before the present evidence had been brought to light, the late Professor Sachs used to teach that Cycads were descended from Ferns of the family Marattiaceæ, and this view may prove to come very near the truth.

Assuming, as I think we safely may, that Marattiaceæ (or Marattiales, if we prefer that more cautious termination, indicating a less close relationship) really existed, as such, in Palæozoic times, it is evident that the Ferns of that period were as highly developed as at the present day, but very differently represented. The Marattiaceous type, the highest of all, was in full strength (the affinities with Pteridosperms are enough to prove the antiquity of the group) and far more widely distributed than now; other living families were for the most part only poorly and doubtfully represented, but there was a great mass of somewhat simply organised Ferns, showing affinities in various directions, and probably representing the common stock from which most of our living tribes of Ferns have diverged. This common stock was itself, however, extremely varied, and only relatively simple.

It may be hoped that future research will enable us to some extent to unravel the very tangled relations between these Palæozoic Fern-groups and the great families of "Seed-bearing Ferns" which had a common origin with them in some more remote past.

CHAPTER VI

THE EVOLUTION OF THE HIGHER SPORE-PLANTS—THE CLUB-MOSSES

THE Fern-series is of the utmost interest from its relation to the main line of descent of the Seed-plants. Apart from this consideration, the past history of the Ferns themselves presents little that is remarkable. The Palæozoic Ferns, in so far as they were true Ferns at all, seem to have been much the same sort of plants as their living representatives. There were lofty Tree-ferns in early days, but so there are still, though they belong to different families.

When we come to the other great classes of the higher Spore-plants there is quite another story to tell. The Club-mosses and Horsetails of our own times are the scattered survivors of what were once great and dominant families of plants. It is only when we have traced their history well back into Palæozoic times that we begin to realise what Spore-plants are capable of.

We will take the Club-mosses (Lycopods) next, partly because of their great importance in the Palæozoic Flora, and partly because they alone, so far as we know, once made a serious attempt to rival the Seed-plants themselves.

We have already given a short outline of the characters of the Lycopods (p. 137). They are still fairly numerous; 478 species of the four genera are enumerated in Baker's *Fernallies*, and the number has considerably increased since. I do not include the small family Psilotaceæ, which seems to have closer affinities with another class (see below, p. 212).

The class is divided into two families—the Lycopodiaceæ with spores of one kind, and the Selaginellaceæ with spores of two kinds. The former need not detain us long, for though very interesting plants in themselves, scarcely anything is known of their history in past times. There are about 100 species of the genus *Lycopodium*, of which five are natives of our own country. Four of these are plants of the moorlands in hilly or mountainous districts, while one (*Lycopodium inundatum*) is frequent in boggy places in lowland countries, as in the New Forest of Hampshire. All have the habit of very large mosses; the stems, which in some species reach a great length, are clothed by numerous small and simple leaves. In most species there are definite cones, which in the Stag's Horn Moss (*L. clavatum*) are very conspicuous and are borne at the end of long, forked stalks. In *L. Selago*, however, there is no special cone, and the sporangia are produced on the ordinary leaves. In all cases a single sporangium is seated on the upper surface of the fertile leaf; with slight variations of position this holds good for all members of the class, whether homosporous or heterosporous. The species, such as *L. Selago*, in which there is no distinction between the sporophylls and the vegetative leaves, are certainly the simpler and may be the more primitive.

While the plant is young the stem has a simple structure, the wood forming a central column with prominent angles, while the bast lies to the outside. Later on, the arrangement becomes more complicated, wood and bast forming alternate bands or becoming intermingled in other ways. It is a point of interest that this peculiar structure often extends to the roots. Each leaf is supplied by a single unbranched vascular strand, a simple arrangement which prevails throughout the whole class, so far as its recent members are concerned. In some of the exotic species of *Lycopodium* the stem is very robust and the leaves of considerable breadth, but the general structure varies but little throughout the genus. The forking of the stem is the characteristic mode of branching.

The sporangium is a comparatively large sac, conspicuous to the naked eye, attached to the base of the fertile leaf on its upper side. It produces an immense number of minute spores ("Lycopodium Powder," used in fireworks). It is only in recent years that the germination of these spores and the prothalli produced from them have become known. One difficulty is that the spores in many cases require to rest an extraordinary time before they will begin to germinate—from three to five years in *Lycopodium Selago* and from six to seven years in the Stag's Horn Moss (*L. clavatum*). Even then the development is astonishingly slow, so that in *L. Selago* it takes from six to eight years and in *L. clavatum* from twelve to fifteen years after the sowing, before the prothalli are mature enough to produce their organs of reproduction. In these native species and some others the prothalli are subterranean, the spores getting washed deep down into the ground before they germinate. There is no chlorophyll, and a very curious point is that the prothalli cannot develop properly unless they are infected by a Fungus, which penetrates their tissue and lives on terms of mutual benefit with its host—a remarkable arrangement, which, however, is not at all uncommon in other cases (see Chapter II).

This does not, however, apply to all Lycopodiums by any means, for in *L. inundatum* and various exotic species, the prothallus grows on the surface of the ground, or on the bark of a tree; it has leaf-like outgrowths, and is green in colour and able to look after its own nutrition like other chlorophyll-containing plants.

In any case the same prothallus produces both kinds of sexual organs. The spermatozoids differ from those of the Fern-group in having, as a rule, only two cilia each. After fertilisation of an egg-cell by a spermatozoid an embryo develops, which remains for a very long time in connection with the prothallus.

Phylloglossum, the only other member of the Lycopods with spores of one kind, is a minute Club-moss growing in Australia and New Zealand; it has a short tuberous stem, a few simple leaves and a little cone. The prothallus is green, like that of a *Lycopodium* of the *L. inundatum* type. The genus is interesting, because the mature plant much resembles the embryonic stage of some species of *Lycopodium*, but whether this means that *Phylloglossum* is a primitive or a reduced form is still an open question.

The family Lycopodiaceæ is a curious and isolated group, which we may reasonably suppose to represent an ancient race. Unfortunately next to nothing is known of its geological history. An Upper Triassic plant, *Naiadita*, has much in common with *Lycopodium*. At present, however, there is no case of a fossil Lycopod in which we can be certain that the spores were all of one kind. Whenever there is definite evidence, the fossil Club-mosses have proved to have spores of two kinds. It is difficult to prove from fossil specimens that only one kind of spore existed, for we cannot distinguish between microspores and the uniform spores of a plant with one kind only, whereas the opposite condition is at once demonstrated as soon as a megaspore is detected.

We now go on to the heterosporous Lycopods, the Selaginellaceæ, a much more satisfactory group from an evolutionary point of view. The two genera, *Selaginella* and *Isoëtes*, are extremely different, and probably not at all closely allied.

Selaginella is very rich in species; Mr. Baker, of Kew, enumerated 334 in the year 1887, and later authors have estimated the number as high as 500. They are graceful plants, many of them very popular in greenhouses; they are often confused with species of *Lycopodiums*, to which they bear a general resemblance, though very different in their life-history, and as a rule more elegant in form.

There is only one British species, *Selaginella spinosa*, which is not uncommon on moors and mountain sides. In this species all the leaves are alike, as in most species of *Lycopodium;* in *Selaginella*, however, it is more common for the leaves to be of two kinds, the stem bearing four rows, two of large leaves and two of small, the smaller being on the upper surface of the stem when it is a creeping one. In many species the branches with their leaves are spread out in one plane, the whole having much the appearance of the compound frond of a Fern.

Some species, such as the Bornean *S. grandis*, stand upright, growing to a height of a couple of feet or more, like little trees, and helping us to realise, in miniature, the aspect of the giant Club-mosses of the Coal-forests. Others, like the *S. Wildenovii* of the Eastern Tropics, attain a much greater size, climbing for twenty feet or more over the branches of more robust plants.

A very general feature in *Selaginella* is the presence of leafless branches, which grow down from the forks of the stem and produce roots as soon as they touch the ground. Whether the root-bearers are themselves of the nature of stems or roots is still disputed; the existence of these intermediate organs is of considerable interest, and characteristic in one form or another of many Lycopods.

The spore-bearing organs, throughout the genus, are in the form of well-defined cones, the fertile leaves often having a different arrangement from that of the ordinary vegetative leaves.

The structure of the stem is extremely diverse in different species. In the simplest forms (*S. spinosa*, for example) there is a single central column of wood, developing from without inwards, and surrounded by bast; in a large number of species, however, the stem contains two or more separate vascular cylinders, of varying form; in some the arrangement becomes highly complex. The leaves, however, have essentially the same simple structure as in *Lycopodium*.

A very characteristic feature is the presence, on the upper side of each leaf, near its base, of a tongue-shaped outgrowth, the *ligule*, which serves to secrete mucilage and keep the young organs moist in the bud. The ligule is constantly present on all leaves, whether vegetative or fertile, both in *Selaginella* and *Isoëtes*, and extends to a large proportion of their Palæozoic allies. It is thus an important character and serves at once to distinguish the Selaginellaceæ and related fossil groups from the homosporous Lycopodiaceæ.

In *Selaginella*, as in all other Lycopods, there is one sporangium to each fertile leaf; in *Selaginella*, however, the sporangium is *axillary*, i. e. seated in the angle between the fertile leaf and the stem, instead of on the leaf itself—a distinction, however, of comparatively small importance.

The two kinds of sporangia are commonly formed in the same cone, those producing the small spores in the upper part of the cone and those producing the large spores in the lower part. It appears that the megasporangia are produced on the *lower* side of the cone, whatever its position. Up to a certain point the two kinds of sporangia develop alike, forming a large number of spore mother-cells. If the small spores are to be produced, each of the mother-cells divides into four, and a great output of microspores is the result. But in order to form the large spores, only one of the mother-cells divides, so that only four megaspores result;

they grow to a great size and fill the sporangium, which is often actually larger than in the former case (see fig. 18). Sometimes, even the four do not all develop and the number of megaspores

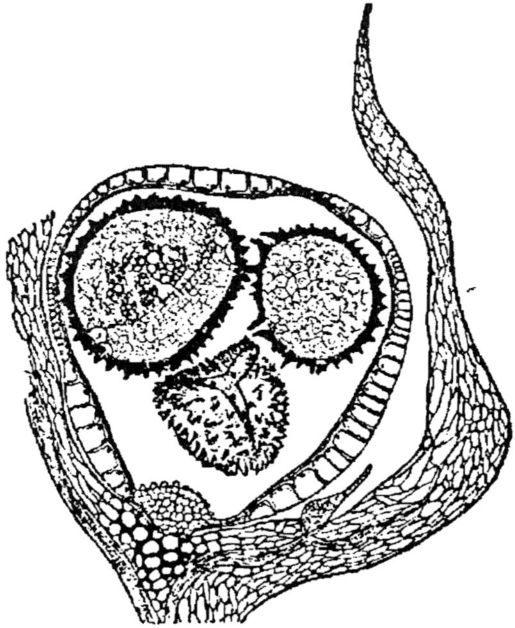

Fig. 18.—*Selaginella.* Section of sporophyll, ligule and megasporangium, showing three megaspores, with the prothallus in two of them. After Miss Lyon. × about 40.

in the sporangium may be reduced to three, two, or even one. The diameter of the megaspores is commonly about ten times that of the microspores, which is equivalent to a proportion of 1000: 1 in bulk; sometimes the difference is greater.

The germination of the megaspores, resulting in the formation of a prothallus, takes place while they are still inside the sporangium (see fig. 18). As a rule the large spores are then shed; the prothallus develops further and bursts through the spore-wall; it does not, however, become green. In the meantime the microspores have likewise been shed, and have each developed a small group of cells, which is practically nothing but an antheridium, and from this a few spermatozoids, each with two cilia, are set free. If water is present, they swim to the archegonia of any female prothallus which is close by, attracted by a secretion of malic acid as in the Ferns. Fertilisation is accomplished and the result is an embryo, which remains enclosed for some time in the megaspore, embedded in the prothallus.

It has already been mentioned that in some species fertilisation takes place inside the sporangium, and an embryo is formed, which may even grow into a young plant, without the megaspore ever being shed at all. This is an extreme case and not altogether normal, though interesting as a parallel to what happens in the regular course of development of the higher Seed-plants. In *Selaginella*, however, there is no seed, for the sporangium remains unaltered.

The heterospory of *Selaginella* is thus extremely

well marked, and this is no wonder, for we have direct evidence that this group has been reproduced in the same way ever since Palæozoic times; indeed, its present highly-developed condition had been fully attained before the close of that period.

The other living genus of heterosporous Lycopods, *Isoëtes*, is totally different in habit from *Selaginella*. There are about fifty species, which are mostly water-plants; one is fairly common in the lakes of our own country, growing entirely under water; another species, one of the two which grow on dry land, extends to the Island of Guernsey, and so is reckoned as a British plant. On the whole, there is not much difference in structure between the aquatic and the terrestrial species; the genus seems to have long hovered between land and water.

In all the species, the stem is extremely stunted, running to breadth more than height. The crowded leaves are long and quill-like. There is no constant distinction between vegetative and fertile leaves, and consequently no formation of a cone.

The stem is deeply grooved on its under surface, the grooves dividing it into two, three, or occasionally four lobes; the roots arise from the sides of the grooves.

The most interesting feature in the anatomy

is that the stem grows indefinitely in thickness by means of cambium, a rare occurrence among living Cryptogams. A considerable amount of secondary wood and bast is added by the cambium, but the cortex is often the part chiefly increased. The whole process is peculiar and differs from the normal secondary growth of a Dicotyledon or Gymnosperm. It is interesting to note that growth in thickness also goes on in the base of the stem of *Selaginella spinosa*. The leaves of *Isoëtes* have the single vascular strand characteristic of the class. As in *Selaginella*, the ligule is a constant feature.

The roots are remarkable in so far as their structure somewhat resembles that of a leaf, the vascular system consisting of a single strand of wood and bast, whereas in typical roots there are two or more strands of each. This peculiarity is also frequent in the roots of *Selaginella*.

The sporangia of *Isoëtes* are unusually large, often quite a quarter of an inch long. Each is seated in a hollow on the upper side of the fertile leaf, between its base and the ligule. There is often a membrane, like the integument of an ovule, which wholly or partially envelops the sporangium. It will be noticed that in *Isoëtes* the sporangium is entirely a product of the leaf, and occupies a considerable part of the leaf-base. It is rather a complex structure, for the cavity is partitioned up by bands of sterile tissue.

A vast number of small spores are formed in the sporangia set apart for them, and even the megaspores are rather numerous, numbering from forty upwards in the megasporangium.

The subsequent development is much the same as in *Selaginella;* the antheridium of the microspore here gives rise to only four spermatozoids, which have numerous cilia, as in the Ferns, instead of only two, as in other Lycopods.

The female prothallus makes no attempt to grow out of the megaspore, and only becomes exposed by the splitting open of the spore-membrane. Fertilisation takes place as in *Selaginella,* an embryo is developed and the cycle is complete.

The two genera of living heterosporous Lycopods are manifestly very different; probably they represent two distinct lines of descent, which have run their course independently ever since the Palæozoic period. The reasons for this view will now become apparent.

The geological history of the *Selaginella* type can be traced back with certainty to the Carboniferous epoch. The Tertiary and Mesozoic records are of little importance, except as attesting the continuous existence of the group, but when we get back to the Palæozoic age the evidence becomes more complete, and several interesting species are known. The name *Selaginellites* has been given to the fossil forms, but some of them can scarcely be distinguished from the living genus. In the Coal Measure species *S. primævus*, which in foliage seems to have combined the characters of different recent forms, definite cones are present in which the sporangia are preserved. The interesting point is that there were four megaspores in each sporangium, exactly as in the living Selaginellas; the plant as a whole might perfectly well belong to that genus, the type thus showing a remarkable persistency from the Palæozoic period to the present time. Other species of *Selaginellites* were more peculiar: *S. Suissei*, from the Upper Coal Measures of France, had a forked stem, with four rows of leaves of two sizes, like so many of the recent species. The cones reached a length of six inches; the sporangia of the upper part contained numerous microspores, while those of the lower part each contained from sixteen to twenty-four megaspores, ten times the diameter of the microspores. The large number of megaspores in a sporangium is interesting and suggests a less advanced condition than that of the modern *Selaginella*. Another Coal Measure species shares this peculiarity and is further remarkable for the fact that the sporangia arose in the axils of ordinary leaves, no special cone being formed. This simple arrangement, though occurring in *Lycopodium*, is not known among recent Selaginellas. From these facts it would appear that in Coal Measure times some members of the group had already reached the same stage of evolution as the modern genus, while others still lagged behind. It is, however, somewhat surprising to find that another Coal Measure plant, with the habit of *Selaginella*, went far beyond any living ally, and produced an organ closely analogous to a true seed. This plant, *Miadesmia*, occurs in a petrified condition in the coal-balls of the Lower Coal Measures of the North of England. Its structure has been studied in detail. The stem is very slender and the leaves appear to have been borne in four rows; their margin is fringed with hairs, and on the upper surface is a remarkably well-developed ligule. The structure of the stem is much like that of one of the simpler species of *Selaginella*.

The fructifications are grouped in a loose cone; on the upper surface of each sporophyll is a sporangium, containing a single megaspore of large size, which occupies the whole cavity of the sporangium, like the embryo-sac in a seed (see fig. 19). The prothallus in the megaspore is sometimes preserved. The megasporangium is enclosed in an envelope or coat, by which it is completely covered; only a narrow opening, which we may call the micropyle, is left, at the far end of the sporangium. The envelope bears a number of long tentacles, which project beyond the micropyle and may have served to detain the microspores. At the base of the sporangium the sporophyll bears a ligule like that on the vegetative leaf. The sporophyll has a broad blade extending far beyond the sporangium on

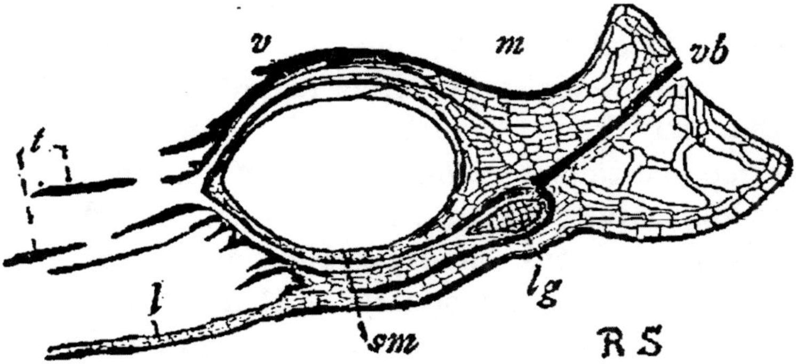

Fig. 19.—*Miadesmia membranacea.* Longitudinal section of "seed"; (*vb*) vascular strand of sporophyll; (*l*) blade of sporophyll; (*lg*) ligule; (*t*) tentacles; (*v*) integument of "seed"; (*sm*) sporangium-wall; (*m*) megaspore. From Scott, Studies. × 30.

either side, so that when the whole was shed the organ was much like a winged seed.

It has been said above that the Palæozoic Lycopods attempted to rival the Seed-plants on their own ground. The *Selaginella*-like plant *Miadesmia* is one of the Club-mosses which came nearest to the formation of a true seed. The envelope or integument, and the single megaspore, retained permanently within the sporangium, are essential characters of a seed. The fertile leaf, however, formed part of the seed-like organ, and the whole was shed together, like the fruit of a Composite.

It is curious to find so great an advance in reproductive methods, so nearly approaching the highest types, in a little inconspicuous Lycopod of Carboniferous age. If, as seems fairly clear, *Miadesmia* was one of the Selaginellaceæ, it rose to a higher level, as regards reproduction, than any known member of its family has done since—certainly higher than any of those now living. Here, as in so many groups, the less advanced forms have persisted while the more advanced have disappeared. It must be remembered that at that time the seed-method of reproduction, though already common, was not so firmly established as it became later. The constant absence of an embryo in Palæozoic seeds suggests that not all the possibilities of the seed-habit had then been realised. Under such circumstances, it may be that there was more scope for experiment in the seed-direction, and more chance of successful competition with the main line of Seed-plants, than at a later time.

As regards the particular conditions which favoured the seed-habit in the case of *Miadesmia*, we are left to conjecture. The specimens of the little plant are very generally associated with one of the large Lycopods, a *Lepidodendron*, which was probably a tree. It has been suggested that the slender and fragile *Miadesmia* may have grown as an epiphyte on the branches of its robust neighbour, just as so many Selaginellas are epiphytic on stronger plants at the present day. If this was so, *Miadesmia*, growing on trees, would have been in much the same difficulty with regard to its reproduction as the tree-Lycopods themselves. It has already been pointed out (p. 123) that the heterosporous method has the disadvantage that the chances are against the microspores and megaspores coming to rest at the same spot, a condition necessary for fertilisation, so that the production of a vast number of microspores is demanded to ensure an occasional meeting. This difficulty became aggravated when the spores were discharged from the branches of tall trees, whether from the cones of the trees themselves, or from those of an epiphyte. If the necessary meeting could be arranged to take place on the parent plant, the matter would no longer be left solely to chance. This advantage was secured by the adoption of the seed-habit, involving the retention of the megaspore within its sporangium. This may have been one of the circumstances favourable to the development of a seed-like organ in heterosporous trees, and also in a plant like *Miadesmia*, if it was an epiphyte on trees.

All this is pure hypothesis, and is only suggested because it may help us to realise that the Palæozoic plants were subject to a struggle for existence like the plants of our own day, and could not afford to indulge in structural elaborations unless they were of real use under the conditions of their life. But it must be acknowledged that nothing is more difficult than to find out why one plant equips itself for the struggle with one device and another attains the same end in quite a different way.

From what has been said it appears that the *Selaginella* line has gone on from Palæozoic times with comparatively little change, except that the highest types have dropped out, owing perhaps to their becoming involved in keener competition with other races of plants.

We have not yet touched the main body of the Palæozoic Lycopods, the Lepidodendreæ. Their relations to living plants are by no means so clear as in the case of the Selaginellaceæ, so we will not attempt in the first instance to trace back their history from living forms, but will first sketch the main outlines of the fossil group, and will then see whether we can connect them at all with any surviving type, or whether it is more probable that they have become altogether extinct.

The characteristic Club-mosses of the Palæozoic period were trees, of which a great number are known. Hundreds of species have been described, from all parts of the world in which Carboniferous rocks have been explored. A remarkable uniformity in the Flora prevailed in those ages; there was little difference between the vegetation of Australia, South Africa or South America, and that of our own country. The tree Lycopods, as their remains show, were extraordinarily abundant, and no doubt formed the chief feature of the swampy forests which made the Coal.

The trees were of great size. Last year a gigantic fossil trunk of a *Lepidodendron* was found prostrate in the Royal Arley Mine, near Bolton. The length of the stem was 114 feet up to the point where it began to branch; the branches and foliage could be traced for another 17 or 18 feet, and even then were incomplete. The trunk had been flattened, owing to pressure or decay, before fossilisation. At the base the thickness was 3 ft. 8 ins. by 1 ft. 8 ins.; at the time of branching, 1 ft. 9 ins. by 4 ins. Allowing for distortion and some loss of substance the diameter, in the natural condition, may have been about 3 feet at the base, and a little over a foot at a height of 114 feet; thus the stem must have been of slender proportions. I am indebted for

Fig. 20.—*Lepidodendron elegans.* Restoration of tree, bearing cones. From Scott, Studies.

these particulars to the discoverer, Mr. James Lomax.

Trunks of the other principal genus, *Sigillaria*, have been found of almost equal height and similar proportions; in some, however, the form was shorter and sturdier; a German *Sigillaria* is described, which was 6 feet in diameter at the base and only 18 feet high, diminishing to a thickness of about a foot at the top.

In habit the tree Club-mosses most resembled *Lycopodium*, but in their mode of reproduction they agreed rather with the Selaginellaceæ. No doubt their true affinities were closer with the latter group, though there was no near relationship to *Selaginella*; as we have seen, the latter type was, for all we know, as ancient as the Lepidodendreæ themselves, and there is nothing directly connecting the two groups.

The Lepidodendreæ (using the name to include *Sigillaria* as well as other less important genera) were already well developed in Devonian times, and extend right through the Carboniferous and Permian, becoming rare and doubtful when we reach the Mesozoic rocks.

In habit the two principal genera, *Lepidodendron* and *Sigillaria*, were different. Both had an upright, usually lofty stem, but *Lepidodendron* was freely branched, and consequently had to some extent the aspect of our familiar forest trees (see fig. 20), while *Sigillaria* branched very sparingly, or in some species not at all, and thus had a peculiar habit, suggesting the Cabbage-tree (*Cordyline*) of New Zealand or the Grass-trees (*Xanthorrhœa*) of Australia. In all cases the branching was of a forked nature, but the two branches of the fork were often developed unequally.

The leaves were numerous, simple and grass-like, attaining a great length (up to a yard) in some species of *Sigillaria*. A very important point is that every leaf had a ligule on its upper surface, near the base, just as in the recent Selaginellaceæ. The fructification was in the form of well-defined cones, often of considerable size; they were usually borne at the ends of the smaller branches, but in *Sigillaria* were commonly produced on the main stems.

The distinction between the two genera *Lepidodendron* and *Sigillaria* depends, in the first instance, on the arrangement of the leaf-bases, which remained on the stem after the leaves themselves had been shed, forming characteristic patterns. The leaf-bases in *Lepidodendron* cover more or less completely the surface of the stem, and are commonly ranged in spiral or oblique rows, while in *Sigillaria* the characteristic arrangement is in vertical lines; in an important section of the genus there are well-marked longitudinal ribs, corresponding to the vertical series of leaf-bases. Of the two, *Lepidodendron* appears to have been the more ancient genus.

The root-system, which seems to have been similar in the two genera, is characteristic. From the base of the trunk four great roots diverged, running horizontally or obliquely through the soil; each of the four forked at least twice; these underground organs reached an immense size; a root measured by Williamson was 37 ft. 4 ins. long and 2 ft. 8 ins. in diameter at its base. The main roots bore numbers of rootlets, which likewise forked, penetrating the soil in all directions.

The internal structure of the stem in the Lepidodendreæ was at first somewhat like that in the simpler species of *Selaginella*, but it varied considerably in the different species, and in most cases became profoundly changed in advancing age by the occurrence of secondary growth. The stem always contains a single vascular cylinder; in some Lepidodendrons the central column of wood is solid; in most species there is a pith, and in some Sigillarias the surrounding zone of wood is more or less completely broken up into separate strands. So far as the primary wood is concerned, the development was always from without inwards, as in most recent Lycopods; in most of the Lepidodendrons, however, and in all known Sigillarias, a considerable thickness of secondary wood was added by a cambium; there was also an extensive development of secondary tissue in the cortex.

At one time the secondary growth of the Lepidodendreæ, and especially of *Sigillaria*, of which it was supposed to be characteristic, misled many botanists into thinking that these plants were Gymnosperms and not Cryptogams. We now know, thanks chiefly to the work of Williamson, that the formation of secondary tissue is by no means peculiar to Seed-plants, but extended, in Palæozoic days, to every group of the higher Cryptogams. In our own time we only find reminiscences of the old capacity for unlimited growth, in *Isoëtes* and a few other cases. No doubt secondary growth was to a great extent correlated with the tree-habit, though it was not limited to trees then, any more than it is among Dicotyledons at the present day. Secondary growth by means of a kind of cambium goes on in some of the larger seaweeds (*Laminaria*, etc.) at the present day, and was even more extensive in a huge Palæozoic Alga, named *Nematophycus*.

Each *Lepidodendron* leaf, as a rule, received a single vascular strand from the stem; the leaf-strands were comparatively small, and passed out without causing any gap in the ring of wood from which they started. This is another important difference between Lycopods and Ferns; in the latter there is almost always a leaf-gap, where the strand of a leaf passes out, because of the greater importance of the vascular supply demanded by the larger leaves of that class.

In some species of *Sigillaria* the vascular strand forked in passing through the cortex, and here we find two parallel strands in the leaf — the only instance among Lycopods.

The leaves are often very well preserved, and it has been found that their stomata were placed in two furrows on the lower surface — a protected position which recalls the arrangements found at the present day in plants of dry climates; Lepidodendreæ certainly did not grow under such conditions, but a similar protection against excessive transpiration is found in the plants of salt-marshes, and this may be a more probable explanation here.

We have called the underground organs "roots" and "rootlets," because these words are intelligible and no doubt are applicable so far as function is concerned. The structure, however, was not very root-like; the main roots had a structure like that of a stem, while the rootlets had the same kind of vascular system which we noticed in the roots of *Isoëtes* — a single strand of wood and bast, more suggestive of a leaf than of a root. There are also some curious complications in the structure of the rootlets, in the way of vascular strands branching off into the cortex, forming a special apparatus for the absorption of water.

There are many instances in Lycopods of structures which do not quite fall under the head of root or of shoot; this may possibly be an indication that the group is in some respects a primitive one, retaining some characters which date from a time when the different organs of the plant-body were not quite sharply differentiated. None the less, the organs in question, and especially those of the fossils, show remarkably perfect adaptation to their functions. The main "roots" of the Lepidodendreæ have been compared to those curious organs, the leafless root-bearing branches of *Selaginella*.

We must now pass on to the organs of reproduction. The sporangia, as we have seen, were always borne in definite cones, the general organisation of which was, broadly speaking, on the same lines as in a recent Lycopod, the genus *Selaginella*, with its two kinds of spores, affording the best parallel, though far from an exact one. Indeed, in some respects the analogy with *Isoëtes* is greater, only in that genus there is no cone differentiated.

There are two grades of organisation in the fructifications of Lepidodendreæ; on the one hand, we have a typical heterosporous mode of reproduction, on about the same level as that of *Isoëtes* and *Selaginella* in the recent Flora. On the other hand, we find reproduction by seed-like organs, such as those which we have already met with in the herbaceous *Miadesmia*. This is a more advanced method than anything we find among living Club-mosses. We will begin with the simpler, heterosporous cones, which seem to have been in the majority, though the seed-like type was also quite common.

The typical cones of *Lepidodendron* had a characteristic structure. The axis of the cone bore a great number of spirally arranged or sometimes whorled sporophylls, closely packed together. Each fertile leaf or sporophyll consisted of a more or less horizontal basal part, terminating in a broader blade, which turned vertically upwards and usually had a shorter downward prolongation as well. The sporangium was attached along its whole length to the upper surface of the horizontal portion of the leaf. Its length was sometimes very considerable — in one specially fine cone the sporangium is almost two centimetres long by five millimetres wide; these are, perhaps, the largest Cryptogamic sporangia known. The ligule was placed at the far end of the sporangium, just below the blade of the sporophyll, showing that the whole of the horizontal part represents a long-drawn-out leaf-base. The insertion of the sporangium along its whole length on the elongated base of the fertile leaf is characteristic of the cones of *Lepidodendron;*

in a species of *Bothrodendron*, a genus in some respects intermediate between *Lepidodendron* and *Sigillaria*, the sporangium was attached by a small base and the sporophyll was correspondingly short. The long base of the sporangium in Lepidodendreæ finds a close parallel in *Isoëtes* alone among living Lycopods.

The sporangia of the fossil genus often contain bands of sterile tissue, running up from the sporophyll, and perhaps helping in the nutrition of the spores. It will be remembered that the same thing occurs in the sporangia of *Isoëtes*.

In some cases the two kinds of sporangia occur in the same cone, the microsporangia above and the megasporangia below, quite like an ordinary *Selaginella*. In other cases, however, it appears that each cone bore one kind of sporangium only. It is impossible to prove that all the species were heterosporous, for sometimes only the small spores have been observed, but the cones agree so closely in structure, so far as the true Lepidodendreæ are concerned, that there is little room for doubt. The microspores were enormously numerous; in the larger sporangia there were a considerable number of megaspores, but in other cases, as in the *Bothrodendron* cone already mentioned, there were only four. The difference in size between the two kinds of spore was very great, often 50 to 1 in diameter, which is equivalent to a proportion of 125,000 to 1 in bulk. The megaspores were commonly about one millimetre in diameter, and in some cones of *Sigillaria* they were double that size; the head of a pin gives an idea of their average dimensions. Megaspores are extremely abundant in some kinds of coal.

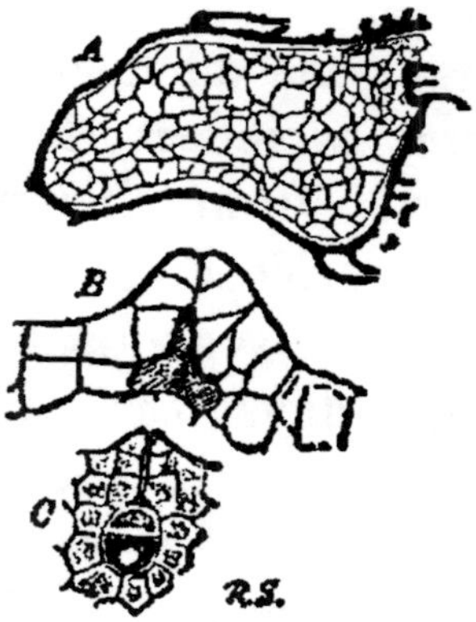

Fig. 21.—(A) *Lepidodendron Veltheimianum.* Megaspore filled with the prothallus, × 25; (B) archegonium from another prothallus of the same plant, × 140; (C) archegonium of *Isoëtes*, for comparison, after Hofmeister, × 200. A and B from drawings by Mrs. D. H. Scott.

Very commonly the megaspores were covered with bristles, among which we often find the microspores entangled; this may have been a means of ensuring that the two kinds of spores germinated together.

In some favourable cases the preservation is so good that the prothallus is still found within the megaspore (see fig. 21, A). It consisted of a uniform, large-celled tissue, except on the side where the megaspore opened; here the cells were smaller, and Mr. W. T. Gordon has recently been so fortunate as to discover a well-preserved archegonium in this position (see fig. 21, B). All the parts are shown — the neck, the canal, and the egg-cell; the agreement with an archegonium of *Selaginella* or *Isoëtes* is remarkably close (see fig. 21, c). Thus we now have full assurance that the process of reproduction in these Palæozoic Lycopods was the same as in the heterosporous members of the group still living. It is interesting to think that this method has gone on unchanged for such immense ages, and to reflect that in *Selaginella* and *Isoëtes* we have, from this point of view, plants of the Palæozoic Flora still flourishing.

The Lycopods, however, have long passed their best days; their most advanced representatives flourished in the Carboniferous epoch, and, so far as we know, have left no descendants. We have already considered one of the "seed-bearing Club-mosses" — the little herbaceous plant *Miadesmia.* The same high grade of reproduction was attained by some of the tree-Lycopods, in which, indeed, the seed-like organs were first discovered. The genus is known by the name of *Lepidocarpon,* and is based solely on the seed-character, for in all other respects the structure of the cone is exactly like that of a *Lepidodendron.* The seeds were known for many years before they were proved to belong to a Lycopod, and were then supposed to be Gymnospermous seeds of the family Cordaiteæ.

In the young condition the arrangement of the parts of the cone is precisely the same as in the cones of *Lepidodendron* already described. The sporophylls are of the usual shape, and on the upper side of each is seated a large sporangium, attached by its under side to the long basal portion of the fertile leaf. A ligule is present in the usual position. But at this stage there is already a striking peculiarity. Almost the whole cavity of the sporangium is occupied by a single megaspore, of enormous dimensions, with its membrane in contact, or nearly so, with the wall of the sporangium. Side by side, however, with the one megaspore which reached maturity, are found three abortive spores, its undeveloped sister-cells; evidently the original number of megaspores in a sporangium was four, of which one had gained the upper hand over the rest. The same thing occurs occasionally in *Selaginella*, and regularly in the heterosporous Water-ferns; analogies are also to be found in the development of the embryo-sac of recent Seed-plants.

At a later stage the seed-like condition was complete. A thick envelope had grown up from the sporophyll, and formed a close investment round the sporangium, only leaving open a narrow crevice along the top; this crevice may be called the micropyle, but it had the form of a long slit, extending the whole length of the sporangium. When cut across, the appearance is altogether that of a seed. In several cases the prothallus has been found well preserved within the megaspore of the "seed," but we do not yet know the structure of the archegonia. The seed-like organ, sporophyll included, was shed as a whole. In some forms it was quite half-an-inch long.

Essentially, the seed-like characters are the same as in *Miadesmia* — the single megaspore, the seed-coat, and the shedding of the organ as a whole, the megaspore being retained within it. In detail, the resemblance to a seed is somewhat less exact in *Lepidocarpon* — the micropyle has an unusual form, and the presence of the three abortive spores shows that the Cryptogamic stage had not been left far behind.

It is quite clear, from many differences in detail between the "seed" of *Lepidocarpon* and that of *Miadesmia*, as well as from the characters of the other organs, that the evolution of the two had been quite independent. Clearly, the need for this reproductive advance had made itself felt in two distinct lines of descent — in the small herbaceous plants of the same stock with *Selaginella*, and in the arborescent Club-mosses which constituted the family Lepidodendreæ. Both these experiments in seed-reproduction seem to have enjoyed a temporary success only — there is no evidence that either was perpetuated or led on to any further advance. The Seed-plants of our own time appear to have been derived from other sources (see Chapters II–IV).

It is an interesting question whether all Lepidodendreæ died out after the Palæozoic period, or whether some of them may have left descendants. As we have seen, they can hardly have given rise to the *Selaginella* line, for the latter type itself goes back to Palæozoic times, and still shows no connection with the tree-Lycopods. There is an evident affinity between the two groups, but we cannot trace them to a common origin.

On the other hand, there are several points in common between the Lepidodendreæ and *Isoëtes.* The sporangium is attached by a long vase to the fertile leaf in just the same way in *Lepidodendron* and in *Isoëtes,* and this is the case in no other groups of Lycopods. The presence of tracts of sterile tissue in the sporangium, and, in certain cases, of an external coat enclosing it, may also be reckoned as common to the fossil and recent groups. Secondary thickening of the stem occurs in both, and the root-structure is similar. On the other hand, nothing could be more different than the habit — tall trees on the other hand, and dwarf-water plants with a flat disc for a stem, on the other. Also, there is no cone in *Isoëtes.*

Can we trace anything of the history of the Lepidodendreæ after Palæozoic times were over? One or two rather doubtful Sigillarias appear in Triassic rocks; in the Upper Triassic of Sweden a large heterosporous cone has been found which might well have belonged to one of the Lepidodendreæ. A curious plant, *Pleuromeia,* from the Triassic rocks of Germany and France, has been regarded as a possible link between the Palæozoic family and *Isoëtes.* The upright stem, a yard or more in height, is marked with leaf-scars, not unlike those of a *Sigillaria,* and terminates below in four short arms, like the four main "roots" of a tree-Lycopod, in a much stunted condition. These four arms are studded with the bases of rootlets. It is evident that we may regard this strange structure as intermediate between the four-rooted base of a Lepidodendroid stem, and the deeply lobed, root-bearing stem of an *Isoëtes.* The idea is certainly tempting, though there are difficulties in carrying out the comparison in detail. Large cones are found with the *Pleuromeia* stems, and no doubt belonged to them. They were heterosporous, for the megaspores have been found. But there is this great difficulty: according to the statements of investigators, the sporangium was seated on the *lower* surface of the fertile leaf, a position quite contrary to what we find in Lycopods. It is impossible to place any reliance on *Pleuromeia* as a connecting link, until this discrepancy has been cleared up.

Some Cretaceous and Tertiary fossils apparently related to *Isoëtes* are known, but they throw no light on its affinities.

We must thus leave open the question of the later history of the Lepidodendreæ; the great Palæozoic family may have become wholly extinct early in Mesozoic times, or may possibly have left a dwindling race of degenerate descendants, which reached their final stage of reduction in the dwarfed plantlets of the amphibious *Isoëtes*. We may sum up, in a few words, what we know of the evolution of the Lycopods generally.

Of the history of the Club-mosses with spores of one kind, we know practically nothing. The characters of the group suggest that it may be an ancient and primitive one, but we have no further evidence.

The *Selaginella* group, among the heterosporous forms, has flourished ever since Carboniferous times, and was then not very different from what it is now. Some of its early members were on the same level as recent forms, others rather less advanced, others again distinctly more advanced, as shown by the possession of organs analogous to seeds.

The main family of Lycopods in Palæozoic times, the Lepidodendreæ, were heterosporous trees, quite distinct from, though related to the contemporary *Selaginella* type. They were a dominant race in their day, when some among them adopted the seed-method of reproduction; they appear to have left few, if any, descendants, *Isoëtes* being the only living genus which may possibly represent the family in a state of extreme reduction.

We have no light of any kind on the *origin* of the Lycopods—they are too ancient for that to be possible, unless the record should be traced much further back than has yet been done. So far as we can follow them, they have always been distinct from any other race of plants, and while varying immensely in stature and elaboration, have always maintained the same general characters, especially as regards the nature of the leaf, and the relation between sporangium and sporophyll.

The Lycopods are an isolated class of plants in the existing Flora, and the extensive records of their past history which we possess do not at present tend to remove them from their isolation.

CHAPTER VII

THE EVOLUTION OF THE HIGHER SPORE-PLANTS —THE HORSETAILS AND THE SPHENOPHYLLS

THE Horsetails are now a very insignificant family compared with the Lycopods, for they number altogether only about twenty species; they are, however, more familiar to the British botanist, for eight species are natives of this country, and some of them are exceedingly common. Every one must know the Field-Horsetail, so common on railway embankments and in neglected fields, and the handsome Great Horsetail, sometimes six feet high, which grows in swampy places, and abounds on wet sea-cliffs. They are mostly striking plants and some almost reach the stature of trees, for a tropical American species is said to grow to a height of from twenty to forty feet, though sometimes aided by climbing on other plants.

All the species have a quite characteristic appearance, owing to their somewhat stiff, jointed stems, and sheathing leaves. Both leaves and branches (where the stem does branch) are in rings or whorls. In some species, as in the two common British Horsetails mentioned above (*Equisetum arvense* and *E. maximum*), there are two kinds of stem above ground, one fertile, bearing the cone, and the other vegetative. The former is short, unbranched, pale in colour, and short-lived; the latter is tall, freely branched, green, and more lasting. In other species, however, such as the Smooth Horsetail (*E. limosum*), common in ponds, there is no such distinction, and the cones are borne on the ordinary green stems. In all cases there are also branched underground stems, penetrating the soil to a great depth, and making the plants most difficult to get rid of when once established. In fact, the Horsetails, though so small a remnant of an ancient family, are still a vigorous race and well able to hold their own; in lakes in the Hebrides, as already mentioned, they are said to be spreading rapidly at the present day, and displacing the aquatic Flowering Plants.

The leaves of the Horsetails are peculiar and much reduced, taking little part in the nutritive work (assimilation of carbon from the carbonic acid of the air) which is ordinarily assigned to leaves. Each ring or whorl of leaves forms a continuous sheath, fitting round the stem; the little teeth projecting from the top of the sheath alone represent the free parts of the leaves.

The usual work of the leaf is in the Horsetails taken over by the finer branches, which are green, or even by the main stem itself. The branches arise at the joints; they are equal in number to the leaves of the whorl, but are placed between them and not in their axils, thus differing from the branches of the higher plants. They grow out through the leaf-sheath, by which they are enclosed when in the bud. The roots grow from the bases of some of the branches of the underground stem; the branches which bear the roots do not usually develope further.

The fructification is always in the form of very conspicuous cones, quite different from the ordinary branches.

Internally, the structure of the stem is much like that of an herbaceous Dicotyledon—there is a ring of separate vascular strands, each of which runs out at the joint into the leaf-sheath, one strand corresponding to each tooth or leaf. The pith is hollow in the main stem and princinal branches; there is also a canal on the inner side of each vascular strand, and often another system of canals in the cortex. Very little wood is developed, except at the joints of the stem, where it connects the vascular strands into a continuous ring.

The roots have essentially the typical structure of the roots of the higher plants, and bear no resemblance to those of the Lycopods.

It may be added that tubers, serving to propagate the plant, are borne, in some species, on the underground branches of the stem.

The cone of a Horsetail has a quite peculiar structure, unlike that of any other living Cryptogam. The rather robust axis bears numerous crowded whorls of shield-like scales, on which the sporangia are seated. Seen from the outside, only the shields are visible, packed so closely together that they become hexagonal from mutual pressure; there are often about twenty of them in each whorl. The shield-like scale consists of a stalk, attached to the axis of the cone at a right angle, and a flat disc which forms the shield. The sporangia are borne in a circle on the under side of the disc, from five to ten in number. Collectively they fill up all the available space between the stalks of the scales.

It will be noticed that I have not called the shield-like scales *sporophylls.* So far as *Equisetum* itself is concerned, there would be no objection to using this word, for though the fertile scales differ much from the vegetative leaves in shape, they agree with them in position and in the way they develope. But when we come to the fossil relatives of our Horsetails, we shall find that there is some doubt whether the fertile scale is really equivalent to an entire leaf, as a sporophyll should be.

The sporangia contain numerous spores, which are all of one kind. A curious point is that each spore has four long bands attached to it, which contract and wrap themselves round the spore when damp, but straighten out when the air is dry; they thus set the spores in motion. This mechanism has something to do with the dispersal of the spores, but botanists have not quite made up their minds how it acts.

The spores, as has been said, are all of one kind. Yet the prothalli are almost always of separate sexes, the male individual being much smaller than the female. The difference, however, does not depend on the spores, but on the conditions under which they grow. If they germinate on a barren soil, such as wet sand, they grow into male prothalli, but if properly supplied with food-materials, a female prothallus is the result. The prothalli are green, and, on the whole, fairly similar to those of the Ferns, as are also the sexual organs of reproduction. The spermatozoids have numerous cilia, just as in Ferns. It is rather interesting that the sexual generation (prothallus) should be so much alike in the two groups, while the asexual plant is so totally different.

The little group of twenty species or so forming the genus *Equisetum* represents the last stage in the history of an ancient family, dating far back into Palæozoic times. We have a very fair knowledge of their history, and the further we trace it back, the more important and the more highly organised do we find the race to have been. In other words, the evolution of the *Equisetum* stock, from Palæozoic times to the present, has been on the whole a steady descent, and not an advance.

The Tertiary species are generally like the larger Horsetails of the present day; the same applies to a certain extent to the later Mesozoic species; there is one form, the Wealden *Equisetites Burchardti*, which had tubers, like our commonest living Horsetails. The genus *Equisetites* had leaf-sheaths like *Equisetum*, from which, indeed, it is often scarcely distinguishable. In the Jurassic rocks larger species are met with, and when we reach the Triassic, gigantic plants of this family occur. The stem of *Equisetites arenaceus*, for example, reached a thickness of eight inches, with 120 leaves in a whorl. In a Rhætic (Upper Triassic) species (*E. scanicus*) from Sweden, M. Halle made the interesting observation that the number of vascular strands in the stem was three times that of the leaves in a whorl; in modern Equisetums these numbers are equal. The excess of vascular strands in the fossil shows that the vascular system was more complex, resembling what we find in the Palæozoic tree-Horsetails. At the same time, *Equisetites* already had the leaf-sheaths of an *Equisetum* and the same form of fructification. M. Halle has made a careful examination of the shield-like scales, in a Rhætic species; they show no important difference from these organs as they exist now, except that the number of sporangia on each scale was greater.

In *Schizoneura*, a characteristic Triassic genus, especially common in the East, the leaf sheaths were split into irregular, leaf-like segments, and in the Rhætic *Neocalamites* the leaves were perfectly free, long and narrow, not forming a sheath at all. In this genus, as in the contemporary *Equisetites* already referred to, the vascular strands were two or three times as numerous as the leaves in a whorl; in fact *Neocalamites*, so far as its characters are known, was almost identical with the Palæozoic genus *Calamites*. In the Triassic Flora the modern and ancient types meet, and there are some signs of transition from the one to the other, though not yet so complete as we could wish. As M. Halle has pointed out, there is no reason to doubt that we have a direct line of descent from the gigantic Triassic *Equisetites* to the Horsetails of our own day. The link between the Triassic plants and the still more highly developed Calamites of the Primary Rocks is not yet perfect, though *Neocalamites* and *Equisetites scanicus* come very near to connecting the two types.

We are, unfortunately, without any petrified specimens of Mesozoic Horsetails, so our knowledge of their structure is confined to the coarser features. In the Palæozoic there is an immense wealth of material of all kinds—petrifactions, impressions, and casts, so that our knowledge of structure is extensive, though there remains the inevitable difficulty of piecing the fragmentary specimens together. However, we can give quite a sufficiently good idea of the group as it was then, for our present purpose.

We may conveniently call the Palæozoic tree-Horsetails "Calamites," from their most important genus *Calamites*. The name was originally given them from a supposed resemblance to Reeds (*Calamus*, a reed), but Suckow, in 1784, for the first time compared them with *Equisetum*. The Calamites thus form a division of the great class Equisetales, of which our modern Horsetails and the Mesozoic *Equisetites* represent the later development. Some of them, if not all, were trees, perhaps not much inferior in height to the giant Lycopods. The tall upright stems appear to have sprung, in many cases, from rhizomes which crept in the mud. The habit seems to have been not altogether unlike that of some of the living Horsetails on an immense scale; the freely branched forms must have been most graceful trees, crowned with a multitude of boughs and twigs, the latter of the utmost delicacy.

The leaves were always whorled, as in the recent forms; they were either quite free from one another or only united at the base; usually, at least in the Coal Measure species, they were of a simple and narrow form. The leaves on the main stems or larger branches, were some inches in length; on the little twigs they were extremely minute, but in crowded whorls, so as collectively to present a large surface to the air and light. The smaller twigs seem to have hung down in many cases, like the branches of a Weeping Willow. The branches were borne at the joints of the stem, in the spaces between the leaves, just as in living Horsetails. In some species the branches were whorled and limited to particular joints of the stem; in others they occurred at all the joints, but often only a single branch developed at each; in others again the branches were few and irregularly scattered. There is evidence that certain branches were periodically shed, and it is probable that this was always the case with the finer twigs; we find an analogy for this in the Larch and the deciduous Swamp Cypress (*Taxodium distichum*) among recent plants.

The cones were highly organized fructifications, more complex, in most cases, than those of any living Spore-plant. They were usually borne at the ends of twigs (see fig. 22), but in the

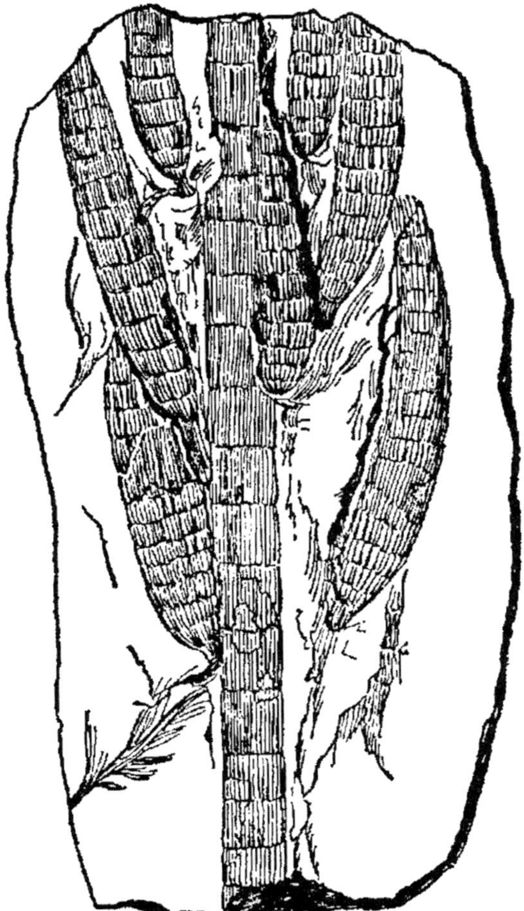

Fig. 22.—*Palæostachya arborescens.* Branch of a Calamite, bearing large cones. After Weiss.

case of some of the largest cones, which reached a length of eight inches, the arrangement was in whorls, on the main stem or a stout branch.

Roots were borne in abundance on the lower part of the stem; in a specimen described by Grand' Eury the stem bore roots up to a height of nine feet from its base.

We will now describe the main outlines of Calamitean structure, for all the organs have been found beautifully preserved; the best specimens come partly from the coal-balls of the English Lower Coal Measures, partly from the Upper Coal Measures of France. Just as in *Equisetum*, the pith was hollow in the main stem, and in all except the finest branches. The commonest Calamite fossils are casts of the hollow pith. Some of these casts are over a foot in diameter, which gives some idea of the size of the tree which had such a pith.

In the young condition, the stem or branch of a Calamite had practically the same structure as an *Equisetum;* there was a ring of separate vascular strands, and every strand had a canal on its inner side. Any botanist looking at a section of such a young Calamite stem would at once call it "a fossil *Equisetum*." But the Calamite did not stop at the *Equisetum* stage. It immediately began to thicken its stem, exactly like a Gymnospermous or Dicotyledonous tree. A broad zone of secondary wood was added, traversed by medullary rays, with a layer of bast to the outside and the cambium between the two; the delicate cells of the latter are in some fortunate cases quite well preserved. Most Calamites are found in this advanced state—it is rare to catch the young stem before secondary growth had begun. The presence of the secondary tissues long misled some of the best botanists as to the true nature of these plants. The good old idea of *Equisetum*-relationship was given up for a time and the Calamites supposed to be Gymnosperms, because it was thought incredible that a Cryptogam should have secondary wood like a Conifer. In fact, the same question arose here as in the case of *Sigillaria* among the Lycopods (see above, p. 179). Increased knowledge of the fructifications, together with the analogy of other groups, has settled the question, and secondary growth in Cryptogams is now just as familiar a conception as it is in the case of Seed-plants.

In *Calamites* itself, the structure of the wood and bast is just about on a level with that in a Conifer, of course with differences in detail. In other cases, especially *Calamodendron*, an Upper Coal Measure genus, the complication was greater, the medullary rays being especially elaborate. Some of the Calamites produced a very thick bark, like the Cork Oak. In most of them the course of the vascular strands was like that already mentioned in some of the Triassic Horsetails; they were more numerous relatively to the leaves than in *Equisetum*.

The structure of the small leaves borne on the twigs has lately been worked out in great detail by Mr. Thomas of Cambridge. They were perfectly organised little leaves, with a single vascular strand, a starch-sheath, a tissue for assimilation, a fibrous strengthening tissue, and an epidermis, which bore stomata on the *upper* side of the leaf alone. This unusual position of the stomata has been explained by the probable hanging position of the whole twig, which would turn the leaves wrong way upwards. The cells of the stoma show certain markings on their walls identical with those observed in recent Equisetums. There are few more perfect cases of the preservation of structural detail than in these leaves.

The roots sprang directly from the stem, not from the bases of branches as in *Equisetum*. They had in some respects the exact structure of Horsetail roots, but reached a much greater size, and, like the stems, produced abundant secondary tissue. An interesting feature is the presence of large air-spaces in the cortex, a characteristic of roots growing in water or very wet soil.

We now come to the structure of the *cones*. As already mentioned, they are, as a rule, more complex than the cone of *Equisetum*. The latter consists of an axis bearing shield-like fertile scales; the Calamite cone, in all the best known cases, consisted of an axis bearing shield-like fertile scales together with sterile, protective bracts. In the commonest type of Calamitean fructification, the arrangement is this: a whorl of fertile scales alternates regularly with a whorl of sterile bracts

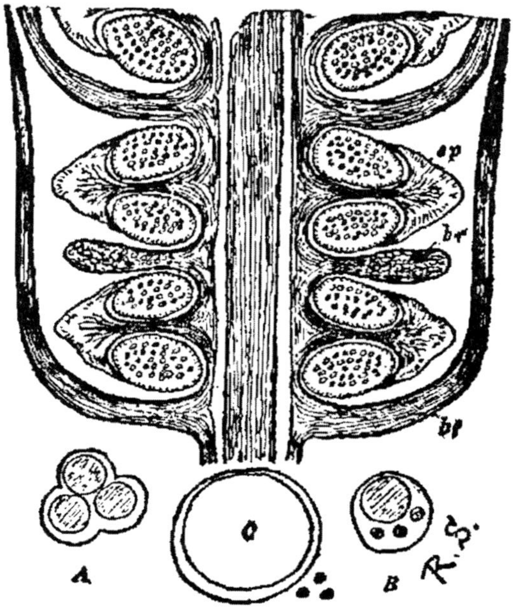

Fig. 23.—Longitudinal section of part of the cone of a Calamite; (*br*) bracts; (*sp*) fertile scales. × 8. (A) Group of four spores, of which three are visible; (B) another group, in which three spores are abortive; (C) megaspore and abortive spores from another species. A, B, C × about 100. From drawings by Mrs. D. H. Scott.

all the way up the cone, and the distances between the whorls are equal, so that the fertile scales come exactly midway between two whorls of sterile bracts and *vice versa* (fig. 23). Usually the bracts in a whorl are about twice as numerous as the fertile scales.

The fertile scales resemble those of a Horsetail in form; they are shield-like, with the sporangia underneath the shield. But in the Calamite the sporangia on each scale are only four in number.

In some French forms, there are further elaborations, the fertile scales being enclosed in chambers formed by outgrowths from the bracts above and below them.

In most of the cones of this type the spores, which are often united in fours (fig. 23, A), are of one kind only, so far as observed, but some of the spores are abortive (fig. 23, B). In one species, however, there are spores of two kinds, in different sporangia on the same cone. The megaspores are about three times the diameter of the microspores, so we have a not very advanced heterosporous condition. It is interesting to find that the large spores are accompanied by abortive spores (see fig. 23, C). This has suggested the idea that in this group of plants we can see how heterospory began. The abortion of some of the spores has eventually allowed the survivors to grow into megaspores. The same result has been produced artificially in recent experiments on some of the Water Ferns.

Anyhow, some of the Calamites were in advance of their successors in this point also, that they had two kinds of spores, though the difference was not so strongly marked as in the heterosporous Lycopods.

In another type of Calamitean cone, the fertile scales, instead of standing midway between the whorls of bracts, sprang from the axil of the bracts below them. This is one of the facts which has suggested a doubt whether the fertile scales were really leaves (sporophylls) or not. It is not very likely that they were branches, but it is quite possible that in both the fructifications in question the scales may have originally been outgrowths from the bracts next below them, and may thus represent parts of leaves, and not entire leaves.

What has been said so far applies to the Coal Measure Calamites, that is to say, to the forms belonging to the later part of the Carboniferous epoch. We do not know so much about the Calamites of the Lower Carboniferous and Devonian, but there are one or two points of interest. The most important of these earlier Calamites is the genus *Archæocalamites*, from Upper Devonian and Lower Carboniferous strata. This plant, which like its successors was a tree, had several peculiar features. The leaves of successive whorls were placed directly above each other, in the same straight line, whereas in the later Calamites and Horsetails, the leaves of each whorl are alternate with those of the next. A more striking difference is in the leaves themselves, which in *Archæocalamites* were long and repeatedly forked, in most marked contrast to the simple little leaves which we usually associate with the Horsetail class. There was, however, one much later Calamite (from the Upper Coal Measures) which retained the forked leaves of earlier days.

The cones of *Archæocalamites* were also remarkable. Instead of whorls of bracts and fertile scales being present in equal numbers, succeeding each other alternately throughout the cone, nearly all the whorls consisted of fertile scales, bracts only occurring, if at all, at long intervals. Thus, in this early type, the arrangement was more like that in an ordinary Horsetail than in other Calamites. The fertile scales themselves, however, bore only four sporangia each, and in that respect were Calamitean. There are some other Palæozoic cones more or less like those of *Equisetum*, though not very much is known about them—at any rate it is clear that in those days there was considerable variety in the fructifications of this group, now so uniform.

There is a Lower Carboniferous Calamite which is interesting from the structure of the wood. In ordinary Calamites, as in *Equisetum*, the wood all lies to the outside of the canal accompanying the vascular strand, and the whole of it was developed from within outwards. In the Lower Carboniferous *Protocalamites*, there is wood on the *inner* side of the canal also, and this part was developed from without inwards. This difference is important for comparison with other groups, especially the *Sphenophyllum* family, to which we shall come presently.

A curious plant, *Pseudobornia*, has been found in the Upper Devonian rocks of Bear Island in the Arctic Ocean. It was a large plant, though scarcely a tree, with a jointed stem and leaves in whorls of about four. The leaves are much more compound even than in *Archæocalamites*, for besides being repeatedly forked, each division is cut into a great number of fine segments; before their connection with the stem was known these leaves were described as the fronds of a Fern. The fructifications, which were in the form of long, loose cones, are imperfectly known; the plant was almost certainly allied to the Calamites and its principal interest lies in the much-divided, Fern-like leaves. Taking *Archæocalamites* also into account, it looks very much as if the ancestors of the Horsetail stock had well-developed compound leaves, which later became reduced in some way, first to the narrow simple leaves of the typical Calamite and then to the mere leaf-sheaths of *Equisetites* and the modern *Equisetum*.

We thus see that the Horsetails of our fields and marshes are the last survivors of a great and ancient race which reached its highest development in Palæozoic times, when it formed one of the leading families of plants. In certain characters, such as complexity of the leaf, the earliest known members were the most elaborate; reduction to greater simplicity in such points is, however, quite consistent with a general advance, and it is probable that the highest level of organisation was reached by the Calamites of the later Palæozoic times—towards the close of the Carboniferous epoch. Species from the Upper Coal Measures certainly show great elaboration, both in the fructification and in the structure of the wood. From that time onwards the downward course began, and its stages can be traced right through the geological record of Mesozoic and Tertiary times, down to our own day. Such records are always imperfect, but the evolution or degeneration of the Horsetail stock can be followed more continuously than that of the Club-mosses. Certain lines must no doubt have died out altogether; for example, the heterosporous Calamites seem to have left no descendants, but still there are good grounds for believing that our existing Horsetails may claim as their ancestors some of the gigantic Calamites of the Palæozoic period.

Our records do not go far enough back to throw any light on the origin of the class Equisetales, but we shall understand their position better after considering another group, wholly fossil, which has some characteristics in common with them, though very distinct in other ways. This is the class of the Sphenophylls, not very extensive according to our present knowledge, but extremely interesting as representing an extinct type of organisation. The genus *Sphenophyllum*, which includes nearly all the members of the class, appears in the Upper Devonian, extends through the whole of the Carboniferous and Permian

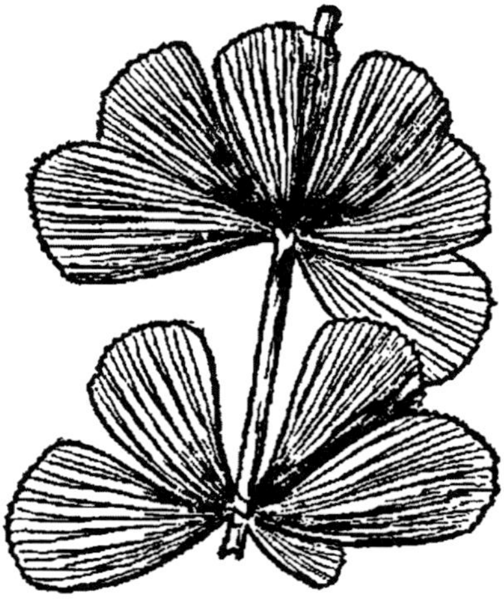

Fig. 24.—*Sphenophyllum*. Fragment of stem, showing two whorls of leaves. After Coemans and Kickx.

epochs, and is lost at the beginning of the Mesozoic. There are, however, some recent plants (the little epiphytic family Psilotaceæ, formerly included under Lycopods) which seem to have some relation to the fossil group.

The species of *Sphenophyllum*—the only genus of which we know enough to give a general description—were slender, graceful plants, very different in habit from the Calamite trees. The stem did not exceed a centimetre in diameter, and even that was an unusual thickness. The jointed, ribbed stems bore whorls of leaves—commonly six in a whorl—at rather long intervals (see fig. 24), the leaves of successive whorls being in the same straight line and not alternating; in this point *Sphenophyllum* resembled *Archæocalamites* and differed from the later Calamites and Horsetails.

The leaves themselves were often wedge-shaped (the name of the genus means "wedge-leaf"), attached by the thin end and traversed by several forking vascular strands. Sometimes the leaves were deeply cut, and often this went so far that they were completely subdivided to the base, so that the six wedge-shaped leaves were replaced by a larger number of quite narrow segments, each of which appears to be a distinct leaf. Both sorts of foliage often occur on the same plant. Branches were borne at the joints, as in Calamites, and roots also.

The fructification in some species formed very definite cones, while in others there was little distinction between the fertile leaves and the ordinary foliage.

The long, slender shoots suggest the idea that the plants may have supported themselves by climbing, like the Bedstraws which scramble over hedges.

The group is an attractive one; the decorative form of the plant is beautifully shown in impressions, and the structure no less perfectly preserved in the petrified specimens. The stem was traversed by a single vascular column; the wood, more or less triangular in section, was solid, with no pith, and its development was from without inwards, as it is in most Lycopods. This applies to the young condition; subsequently a zone of secondary wood and bast was added by means of a cambium, and at the same time a bark of some thickness was formed, displacing the original cortex. The roots grew in thickness in the same way. The leaves of the wedge-like form, with their numerous vascular strands, somewhat resembled the leaflets of a Fern (fig. 24); when cut into narrow segments the structure was simpler, with one strand only in each.

The fructification was very characteristic. The cones, when well defined, bore a general resemblance to those of the Calamites. The sporangia were borne on organs sometimes much like the fertile scales of that family, but there was much more variety. In a very complex cone of the Sphenophyll family named *Cheirostrobus*, from the Lower Carboniferous rocks of Scotland, the sporangium-bearing organs have four sporangia each, and are almost exactly like those of a Calamite; the arrangement, however is different. The cone was a large one—four centimetres in diameter and ten or more in lnegth. It bore numerous crowded whorls of compound sporophylls, ten to twelve in each whorl. The sporophyll was divided into six parts, three above and three below, united into one at the base. The three lower segments were sterile, like the bracts of a Calamite, while the three upper were the sporangium-bearing organs, the shield-like fertile scales. The characteristic point is that they here form part of a compound sporophyll, of which the bracts are also part.

In various species of *Sphenophyllum* the same arrangement as in *Cheirostrobus* exists, with important differences of detail. Thus in the cone of *Sphenophyllum Romeri* each fertile scale bore only two sporangia; this was also the case in *S. fertile*, but in the latter there were no sterile bracts, both the upper and lower parts of the sporophyll dividing up into sporangium-bearing organs. In *S. Dawsoni* we get the sterile bracts again, but here there was only a single sporangium on each sporangiophore (a convenient abbreviation of "sporangium-bearing organ"); two of the latter were seated on the upper surface of each bract (fig. 25).

In all these forms the sporophyll was divided into a lower and upper portion, each being often subdivided. Usually the upper portion consisted of sporangiophores and the lower of sterile bracts;

in *Sphenophyllum fertile* both alike were fertile, developing as sporangium-bearing organs.

There can be no doubt that the Sphenophylls were related to the Calamites. Apart from general resemblances, such as the jointed stems and whorled leaves, the structure of the sporangia and the organs which bore them was practically identical in members of the two groups. It should be mentioned that in the Sphenophylls, so far as we know, all the spores were of one kind, but sometimes there was considerable variation in the size of the spores, and abortive spores were present. It has been suggested that a tendency towards the formation of two kinds of spores may thus be indicated (see above, p. 207).

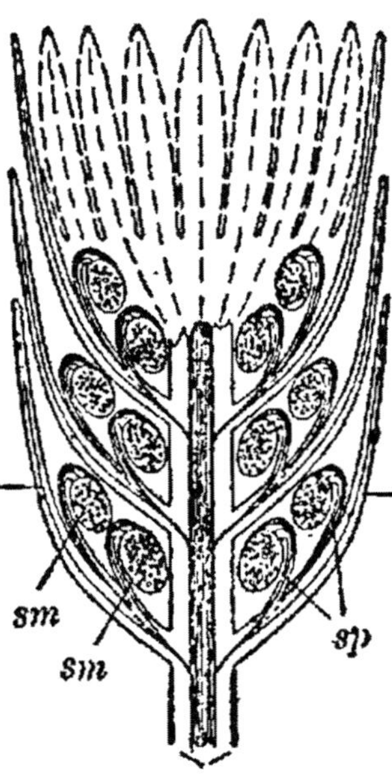

Fig. 25. — *Sphenophyllum Dawsoni*. Diagram of cone, in longitudinal section; (*ax*) axis; (*br*) bracts, some in section, others seen in surface-view from the inside; (*sp*) sporangiophores; (*sm*) sporangia. From Scott, Studies.

There are two important differences (apart from size) between the Sphenophylls and the Calamites.

One is in the structure of the stem, the other in the position of the sporangium-bearing organs.

In Sphenophylls the primary wood was solid and developed from without inwards; in Calamites there was a pith, and the whole of the wood developed from within outwards. The Lower Carboniferous *Protocalamites* (see above, p. 209) helps to bridge this gap, for there a certain part of the wood developed from without inwards, encroaching on the pith, though not to any great extent.

In Sphenophylls the sporangium-bearing organs were always parts of the sporophyll; the bracts, if present, were sterile segments of the same sporophyll. In most Calamites the sporangium-bearing organs were independent of the bracts; we might suppose that each sporangiophore (shield-like scale) represented an entire sporophyll.

There are, however, some reasons, connected with the course of the vascular strands, for supposing that where bracts were present in the Calamites, the sporangiophores had originally been connected with them, *i. e.* with the whorl of bracts next below, so that in these cases the arrangement may have once been the same as in the Sphenophylls. On the other hand, it is not necessary to press this interpretation in every case. In the Ferns we see that in some forms a part of a leaf is set apart to bear the sporangia, as in the Royal Fern (*Osmunda regalis*), while in others a whole leaf is modified for this purpose (*e.g.* the Ostrich Fern, *Struthiopteris germanica*). It may well be that the same differences existed among Equisetales and their allies, the sporangium-bearing organ sometimes representing an entire leaf and sometimes a part of one.

The leaves in Sphenophylls are better developed than in most of the Horsetail class, and suggest those of Ferns. This was the case also, and in a more striking degree, with the leaves of *Pseudobornia*, a genus which may be allied to the Sphenophylls as well as to the Calamites. It has been suggested that both groups may have had a common origin from primitive Ferns, a suggestion which is to a certain extent supported by the characters of the prothalli and sexual organs in *Equisetum*. But any such connection, if real, must lie very far back, and is at present merely a subject for speculation. All that the evidence really indicates is that the Sphenophylls and Equisetales are branches of the same stock, and that they started from plants with large and well-developed leaves.

The history of the common stock is still unknown. Perhaps we may learn something about it when the Devonian Flora has been more thoroughly investigated. It might help us a good deal if we knew the internal structure of *Pseudobornia*, which may quite possibly connect the Sphenophylls with the Calamites. However, it seems to be established that species of the genus *Sphenophyllum* already existed in the Upper Devonian, so we shall have to go further back than that if we are ever to find their ancestors.

The problem is extremely interesting, for while the Sphenophylls approach the Ferns in their foliage, in their anatomy they have something in common with the simpler Lycopods. It may be that we are here on the track of a very early race, in which the characters of the different classes of vascular plants were still to some extent combined, or rather had not yet separated out. If that is so, the Sphenophylls must represent the last survivors of a great group of primitive (but not necessarily simple) Spore-plants, which possibly reached its maximum in pre-Devonian times.

CHAPTER VIII

CONCLUSION

THE student of the evolution of plants is placed under a great disadvantage by the shortness of the fossil record; in this respect he is decidedly worse off than the zoologist, for fossil animals go further back than fossil plants. At present we cannot for our purposes get beyond the Devonian; previous records are still doubtful as regards Vascular Plants; they are of some value for the evolution of certain groups of Seaweeds, but these facts do not in any way connect with the later Palæozoic history, and have no bearing on the descent of the higher plants.

A remark of Darwin's applies with special force to fossil plants: "The geological record, at all times imperfect, does not extend far enough back, to show with unmistakable clearness that within the known history of the world organisation has largely advanced" (*Origin of Species*, p. 308).

In the Devonian period the Vegetable Kingdom had already reached a very high level of organisation. The following main groups of the higher plants are known to have existed at that time—

Cordaiteæ,	Pteridosperms,
Ferns,	Lycopods,
Horsetails,	Sphenophylls.

As regards the Seed-plants, we have found such strong evidence of affinity between the Pteridosperms and the Ferns as to leave no reasonable doubt that they were derived from a common stock; the Cordaiteæ have so much in common with the Pteridosperms that they must ultimately have had the same origin, though their connection with the Fern line of descent was more remote.

The Horsetails and Sphenophylls show unmistakable signs of having sprung from a common source.

Beyond this we cannot go with any safety; we are left with three great races of Vascular Plants: the Ferns, the Lycopods and the joint race of the Horsetails and Sphenophylls.

It will be noticed that the Lycopods run an isolated course all through; their slight convergence with the Sphenophylls in the stem-structure, is scarcely enough to establish any connection.

As we saw in the last chapter, the Sphenophylls may perhaps represent the last remains of an ancient stock, common not only to the Horsetail line but also to other races of the higher Cryptogams. There are so many points of agreement between all these classes that we are led to believe that they must have originally sprung from one ancestral group. This, however, is pure hypothesis. When we get back to the Devonian period, a veil falls, and all the earlier course of evolution (immensely the greater part of the whole history) remains hidden. Scientific men, however, are not always deterred from theory by the absence of facts, and there has been much controversy on the origin of the higher Cryptogams. It is at this point that we enter the region of speculation, while, as we have seen in Chapter IV, quite good evidence for the origin of Seed-plants from the Cryptogams is available.

The question of the evolution of the higher Spore-plants has turned chiefly on the origin of the alternation of generations which is so universal and striking a feature in their development. A similar alternation of sexual and asexual phases exists in the class of the Mosses and Liverworts, but here the functions of the two generations are differently allotted. In the Ferns and their allies, the asexual generation does all the important vegetative work, and is what we call the *plant;* the sexual generation (prothallus) is comparatively unimportant except as the bearer of the sexual cells. In the Moss-series, on the other hand, the *plant,* which does the vegetative work, is the sexual generation; the asexual generation is a *fruit,* which has little more to do than to produce the spores. Thus the *Moss-plant* corresponds to the *Fern-prothallus,* and the *Moss-fruit* to the *Fern-plant.*

The Mosses and Liverworts (together called Bryophytes) are much simpler plants than the Ferns and their allies. Mosses, for example, have no vascular system, though some of them have conducting tissues which partly answer the same purpose. Among the Liverworts, especially, there are some very simple plants, which may be compared to a Fern-prothallus. When, of two related groups of organisms, one is decidedly simpler than the other, there is a tendency to assume that the simpler group represents the ancestry of the more complex group. Thus many botanists have believed, either that the ancestors of the Fern-group were actually Bryophytes, or that they at least resembled Bryophytes in their life-history. This belief involved the theory that the Moss-fruit is an older type of the asexual generation than the Fern-plant, and great efforts have been made to show how the highly organised plant of the vascular Cryptogams might have been evolved by the elaboration of a fruit like that of a Bryophyte. A vast amount of most valuable research has been carried on under the influence of this hypothesis, which has thus played a useful part in science. It is very doubtful, however, whether any light has thus been thrown on the actual course of evolution.

The fossil record, so far as it goes, by no means supports the theory; the oldest known vascular Cryptogams do not show the slightest approach to a Bryophytic fruit or sporogonium or anything of that nature (Sporogonium is the technical word for a spore-bearing fruit developed from an egg-cell, as in Bryophytes). The large-leaved types, the most remote from the supposed sporogonium origin, were more prevalent in the earliest known Palæozoic Flora than they are now. Small-leaved plants existed, namely, the Lycopods; they were mostly trees, and such herbaceous forms as are known were as highly organised as their living successors.

Neither is there any evidence of the superior antiquity of the Bryophytic type. There are only one or two doubtful fossils of Palæozoic age which might be referred to Mosses or Liverworts, whereas, as we have seen, the Vascular Cryptogams were enormously developed in the earliest land-floras of which we have any record.

Negative evidence can never be conclusive, but it is clear that the idea of the superior primitiveness and antiquity of plants of the Bryophyte type remains a pure assumption, and receives no support from our knowledge of ancient vegetation.

Neither among living nor fossil plants has any indication of a structure intermediate between the plant of a Vascular Cryptogam and the fruit of a Bryophyte ever been discovered. The elaborate structural and developmental comparison of the two bodies, worked out with the utmost labour and ingenuity, has brought them no nearer together; the plant remains a plant and the fruit a fruit, as before.

If, then, the theory that the asexual plant of the higher Cryptogams was derived from a sporogonium is unsupported by evidence, what other theory is tenable? Obviously, the simple one that the vegetative plant has always been a vegetative plant—that the Fern with its stem and leaves corresponds to the Seaweed, in which stem and leaf are not yet differentiated, the whole plant being a *thallus.* On this view, the life-history which we now find in the Ferns consists essentially in the regular alternation of an individual bearing sexual organs and an individual bearing sporangia, the two individuals having been equivalent to start with, though one has come to be more highly developed than the other. There are excellent analogies for this among the Seaweeds, especially in *Dictyota,* a brown Alga, common on our coasts, with a forked, strap-shaped thallus. Here there are sexual organs and asexual spores, always borne on different plants. The sexual and asexual individuals alternate regularly. The fertilised egg-cell grows up into a spore-bearing plant, and when the spores of the latter germinate they produce the sexual plant again, and so on. The alternation is just as regular as in the Fern, but the sexual and asexual individuals are exactly alike. There is no reason why they should not be, for both spend their lives in the sea, under precisely the same conditions.

In a Fern it is different; here the prothallus is tied down to a semi-aquatic mode of life by the necessities of fertilisation, which demand the presence of water; the asexual generation, with its spores dispersed in air, is under no such limitation, and can develop as a thorough-going land-plant, with all the high organisation, vascular system, mechanical supporting system and so on, which life in air demands.

While the asexual individual advanced, the prothallus probably became reduced, for the land-plant had to make the most of its land-phase and to economise on the less adaptable, water-bound prothallus.

On this theory, then, the sexual prothallus and the asexual plant are both alike derived from a thallus, and may once have been perfectly similar to each other; the one has gone up and the other down.

The idea of stem and leaf having originated from the differentiated branches of a thallus is quite familiar. Examples of transitional forms between a thallus and a leafy stem are quite well known among Liverworts, and also occur in some of the higher Seaweeds. These are only analogies, it is true, but they give the thallus-theory a great advantage over the sporogonium-theory; we know plenty of intermediate stages between a thallus and a leafy stem, but no one ever saw an intermediate stage between a sporogonium and a leafy stem.

We may now ask, how does the thallus-theory fit in with the fossil record? The record is consistent with the theory but affords it no decisive support. In some of the early Ferns there is a certain resemblance in structure between leaf and stem; it may even be difficult to tell from the anatomy which organ we have before us. It is doubtful, however, whether this occasional resemblance is really a sign that the plants were primitive and their organs not yet fully differentiated. The resemblance between leaf and stem may only have been due to reduction in both, or to some special alteration in the fertile fronds.

Some of the forked forms of leaf met with in early Palæozoic Ferns and in *Pseudobornia* rather suggest a thallus; in a general way the prevalence of large compound fronds in early times is consistent with the theory, for one can easily see how certain branches of a much-divided thallus might have become changed into compound leaves; precisely the same thing is seen in certain Liverworts at the present day, though of course this in the sexual and not the asexual generation. Though the fossil records do not go far enough back to give any strong support to either theory, they agree quite well with the origin of the leafy plant from a thallus, while there is no Palæozoic plant known which we could well suppose to have been constructed out of a moss-fruit or any kind of sporogonium.

It is at any rate safe to say that there is no evidence, fossil or otherwise, for the evolution of the higher Cryptogams from Bryophyta or any plants at all like them. It is more probable that they came direct from plants which were rather of the nature of Algæ; this view, however, is a pure hypothesis and must not be considered as by any means on the same level with the conclusions as to the evolution of the various groups dealt with in the preceding chapters, for these conclusions, whether proved or not, are all based on genuine historical evidence.

The question of the evolution of the Mosses and Liverworts does not fall within our scheme; at present we really have no clue. Far from helping us to understand the evolution of the higher plants, the Bryophytes are themselves about the greatest puzzle of all. It now seems fairly clear that the simpler forms of Liverworts are reduced from more complex forms, and it is not improbable that the whole class of Bryophytes may have owed its origin to reduction from some higher group, of which we know nothing except that it must have had some degree of kinship with the Vascular Cryptogams. The presence of typical stomata, like those of the higher plants, on the fruit of many Mosses and a few Liverworts is a character pointing to their origin from some highly-organised group of plants. The absence of any geological evidence, however, makes the question of the origin of Bryophytes an almost hopeless one.

We must now leave the consideration of special lines of evolution, and endeavour, as briefly as possible, to draw one or two conclusions of a more general kind from the facts before us.

The first and most obvious result of our inquiries is to prove the enormous antiquity of highly-organised plants. If a botanist were set to examine, without prejudice, the structure of those Devonian plants which have come down to us in a fit state for such investigation, it would probably never occur to him that they were any simpler than plants of the present day; he would find them different in many ways, but about on the same general level of organisation. Within the period from the Devonian age to our own time organisation is not shown to have "largely advanced," though there have been many changes. It is not contended that there has been *no* advance; the special adaptations of the Flowering

Plants to Insect life and in other ways show progress in many directions, corresponding to increased complexity in the conditions of life. It must be borne in mind, however, that we know very little as yet about such special adaptations among plants of earlier periods

A very curious point in the geological history of plants is the uniform character of the Flora all over the world during long periods. Professor Seward has recently said: "In dealing with the relation of the Yorkshire Jurassic Flora with that of other parts of the world, it is remarkable to find that almost precisely similar plants to those occurring in the local rocks also are found embedded in strata of about the same age at places so far distant as Bornholm, Poland, Turkestan, Siberia, Korea, Japan, Franz Josef Land, Spitzbergen, Greenland, America, India and Australia."

A similar statement would hold good for the earlier Palæozoic Floras, though at the close of this period there was more distinction in the vegetation, at any rate of the Northern and Southern hemispheres.

The uniform Flora of the world at the periods mentioned is usually regarded as a proof of comparative uniformity of climate. It is possible, however, that the explanation may lie in the greater adaptability of the plants themselves. Certain plants at the present day make themselves at home in almost any climate, and are consequently spread all over the world. This is the case, for example, with various Ferns; the common Bracken extends through all temperate and many tropical regions, in both hemispheres. The Bladder-fern (*Cystopteris fragilis*) also has an extremely wide distribution, extending from the Arctic regions to New Zealand and Tasmania. It may well be that in former ages this adaptability and comparative indifference to climate, now limited to a few cosmopolitan species, was common to most plants, and may account for the world-wide distribution of so many Palæozoic and Mesozoic forms. If so, the special accommodation of particular species to particular climates and environments may be a characteristic of the later Floras, and so far there may have been an advance in exact but limited adaptation. Plants, like their investigators, have become specialists!

It was held by Darwin that "the period during which each species underwent modification, though long as measured by years, was probably short in comparison with that during which it remained without undergoing any change" (*Origin of Species*, p. 279).

On this view there have been long periods of stability, when the conditions of life remained fairly constant, and plants or animals were so far adjusted to them that no great changes were called for. Under such circumstances Natural Selection would simply maintain the existing standard of adaptation, without, on the whole, favouring any new forms that might arise. It is probable that we are now living in a stable period of this kind, so far as our own part of the world is concerned. It appears that there has been very little change in European plants since glacial or even pre-glacial times. There has been a good deal of migration, but the species themselves have, it seems, altered very little.

The geological record shows, on a greater scale, that times of comparative constancy have alternated with intervals of apparently rapid change. Thus, throughout all the middle part of the Mesozoic period, from the Upper Triassic to the beginning of the Cretaceous, the Flora remained much the same; after that the Angiosperms appeared and a complete transformation soon took place. During the Upper Carboniferous age the changes in the Flora were small compared with those that ensued about late Permian times.

It is evident that at any period when species remained nearly constant for a long while, they must have been quite well adapted to the conditions of life that then prevailed; otherwise new variations would have gained the upper hand, and species would have changed. We may therefore safely conclude that during all such periods adaptation to the existing circumstances has been for practical purposes complete. If the plants of later periods seem to us more perfect than their predecessors, it is because they are adapted to conditions with which we are more familiar, and which are also, in some respects, more complex. However far back we may follow the fossil records, we shall never meet with badly-adapted plants—only with plants adapted to different conditions from those of the present. Certain conditions have remained constant all through, and where this is the case, we find just the same constructions in Palæozoic as in recent plants.

The long, rather narrow leaves of the Cordaiteæ were fitted to resist mechanical strains in the same way as Monocotyledonous leaves of similar shape at the present day. In both there are parallel ribs of strong fibrous tissue running through the leaf near the upper and lower surface, *i. e.* in the position where they afford the most resistance to bending with the least expenditure of material. In the outer cortex of the stem of *Lyginodendron* and many other Palæozoic plants, there are bands of fibres united to form a network, the meshes of which are filled up with the softer cellular tissue. This formed an admirable means of support, placed at the outside of the cylindrical stem, where it ought to be placed on engineering principles. The plan worked very well as long as there was not too much increase in thickness of the wood and bast inside. In some Sigillarias and Lepidodendrons, however, the growth in thickness was considerable, so as eventually to burst the outer framework, and here we find that the supporting tissues were constantly renewed from the secondary cortex as growth went on. Thus at all stages the necessary mechanical strength was provided just where it was most wanted, on the outside of the trunk.

The subject of the arrangement of mechanical tissue in ancient plants is an interesting one, because strains due to gravity have been about the same in all ages, so that we can easily compare the corresponding adaptations in the oldest cases known with those in plants now living.

In all respects, wherever we have the means of forming an opinion, the organs of the oldest fossil plants prove to have been just as well fitted to their work as in plants of our own times. The leaves of *Sigillaria* and *Lepidodendron* had two deep grooves on the under side, in which the stomata were sheltered; these were further protected by hairs partly closing the grooves. Quite similar devices are common in leaves of living plants which are exposed to drought, the advantage gained being to check the too rapid loss of water by transpiration through the stomata or pores. Plants growing in salt swamps also have to limit their transpiration, to avoid having to take in too much salt water. Probably this was the explanation in the case of the Palæozoic Lycopods.

The system of secondary thickening, *i. e.* the power of developing unlimited wood and bast by the activity of a layer of dividing cells, the cambium, adding new wood to the inside and new bast to the outside, is an excellent adaptation for enabling plants to grow into trees. We are accustomed to associate secondary growth chiefly with highly-organised Seed-plants, the Conifers and the Dicotyledons. In Palæozoic times, we find that all the groups of Vascular Cryptogams had adopted secondary growth—it is no new advance, but has existed from the earliest times, whenever there was need for it.

Sometimes quite special devices connected with the secondary tissues were anticipated in the Palæozoic Flora. In some Conifers, at the present day, there are water-conducting vessels in the pith-rays which run out through the wood between pith and bast. The vessels in the rays serve as a water-connection between the inner and outer layers of wood. Just the same mechanism isfound in the wood of *Lepidodendron;* the two parallel structures must have been evolved quite independently.

The Seed in its earliest known forms was an elaborate structure, much more so, in fact, than in most recent plants. The pollen-chamber and the very complete system of vascular strands are characteristic features. That the mechanism for the reception of the pollen-grains worked well is shown by the great number of grains (100 or more) sometimes found within the pollen-chamber. Whether so many could have been simply brought by the wind or whether insects were already engaged as pollen-carriers is an open question. The complexity of Palæozoic seeds was almost certainly connected with the spermatozoid method of fertilisation, which doubtless then prevailed, and which still lingers in the Cycads and the Maidenhair-tree. Room had to be provided for the movements of the actively swimming sperms discharged from the pollen-grains. Water for them to swim in was also a necessity, and most likely the vascular system of the seed served to supply it. In later days the evolution of the pollen-tube allowed of a considerable simplification of the seed, or rather of the ovule. But under the circumstances which prevailed in early periods the Palæozoic seed was evidently a very perfect piece of mechanism.

The subject of adaptation in fossil plants is much too wide to be pursued further here. Their study has made an interesting contribution to the theory of Evolution in affording abundant evidence that at all known periods, adaptation, under the existing conditions, has always been as perfect as it is now.

In limiting this little book to the subject of the evolution of the higher plants we have had the advantage of dealing with those groups for which historical evidence, from the fossil record, is available. As regards the rest of the vegetable kingdom—the Bryophytes, the Algæ and the Fungi—there is as yet comparatively little help from fossils; for the most part the evolutionist has here to rely on the comparison of living forms. Even in the more favourable field which we have chosen the work is difficult enough, and its results still tentative and provisional.

The whole problem of Descent is in fact extraordinarily complex, and we are now only at the beginning of the investigation. Evolution, during the periods to which our records extend, proves to have been by no means a regular advance from the simple to the complex. Very often, indeed, the reverse has been the case. The Horsetails and Club-mosses of the present day are very much simpler than those of the Palæozoic period; the Cycads are now much less highly organised than their Mesozoic relations. What the record really shows is a succession of dominant groups, each of which reached a very high development in its time, and then, as the conditions changed, fell into the background, some new family springing up to take its place.

The leading Cryptogamic races of the Palæozoic age, the gigantic Club-Mosses and Horsetails, had lost their importance by Mesozoic times. The great Palæozoic class of the Seed-Ferns was replaced all over the world by the Cycadophytes, while at the same time true Ferns seem to have become more prominent than before. The Cordaiteæ was succeeded by the Coniferæ and by the family of the Maidenhair-trees. When Cretaceous times came, the dominant Cycadophytes soon dwindled away and their place was at once taken by the Angiosperms, probably an offshoot of their own stock. This immense change can be expressed in a word, but we have no idea as yet how the revolution was effected, nor how the infinite variety of Flowering Plants was developed. Beyond the probable connection with the Cycads, we know practically nothing as yet of the course of evolution of Angiosperms, though there are good grounds for believing that the Dicotyledonous type is more primitive than the Monocotyledonous. There is an immense amount of material for the geological history of the Flowering Plants, and we cannot doubt that ultimately its investigation will yield important results.

The present position of the subject of the evolution of the higher plants is an interesting one; we have gained an idea of the relation of the main groups, and know something of their history, but at every point more evidence is urgently needed and all the details still remain to be filled in.

GLOSSARY

Albuminous.—Containing a nutritive tissue (endosperm) when ripe; used of seeds.

Algae.—Seaweeds or fresh-water weeds, containing chlorophyll and reproduced by spores.

Antheridium.—The organ in which the male cells are produced in Spore-plants.

Archegonium.—The organ in which the female cell is produced in the higher Spore-plants and in Gymnosperms.

Bacteria.—Minute unicellular plants, the chief agents in disease and decay.

Bast.—The tissue which serves to convey the food-substances formed in the leaf.

Calyx.—The outermost set of leaves in a flower; usually green.

Cambium.—The tissue from which new wood and bast are formed in secondary growth.

Carpel.—A fertile leaf which bears or encloses the young seeds.

Cilia.—Threads of protoplasm by means of which active male cells and spores move.

Corolla.—The set of leaves in a flower next within the calyx; usually of a bright colour.

Cotyledon.—A seed-leaf; the first leaf, or one of the first two leaves produced by the embryo.

Cryptogam.—A plant reproduced by spores, not by seeds.

Egg-cell.—The female cell which, when fertilised, produces the embryo.

Embryo.—The young plant produced from the fertilised egg-cell.

Embryo-sac.—That cell of the ovule in which the egg-cell and endosperm are formed; the megaspore.

Endosperm.—A tissue formed in the embryo-sac to feed the embryo.

Epiphyte.—A plant which grows on another plant without feeding upon it; like moss on trees.

Exalbuminous.—Containing little or no endosperm when ripe; used of seeds.

Fertilisation.—The union of the male cell with the female or egg-cell.

Fungi.—Spore-plants without chlorophyll, parasitic on other organisms, or growing on the products of their decay.

Gynæceum.—The part of a flower in which the ovules are produced.

Heterosporous.—Producing two kinds of spores: microspores (male) and megaspores (female).

Homosporous.—Producing only one kind of spore.

Hypogynous.—Inserted below the pistil; used of the stamens and corolla.

Indusium.—A membranous coat enclosing a sporangium or group of sporangia in Ferns.

Integument.—A coat or envelope of an ovule or seed.

Lichen.—A plant made up of an Alga and a Fungus, living in close association.

Medullary ray.—A vertical plate of cells, running through wood and bast in the direction from pith to cortex.

Megasporangium.—A sporangium containing megaspores.

Megaspore.—A large spore, producing the female prothallus.

Micropyle.—The opening in the coats of an ovule, through which the pollen-grain or pollen-tube enters.

Microsporangium.—A sporangium containing microspores.

Microspore.—A small spore producing the male prothallus, which is almost reduced to an antheridium.

Nucellus.—The central part of an ovule or seed, surrounded by the integument, and containing the embryo-sac.

Nucleus.—That part of the living body of a cell by which its development and multiplication are controlled.

Ovary.—The part of the pistil in which the ovules are contained.

Ovule.—The young seed before fertilisation.

Palæontology.—The study of fossil plants and animals.

Parasite.—A plant which grows on another plant and feeds upon it; like Dodder on Furze.

Perianth.—The envelope of a flower, including calyx and corolla.

Pericarp.—The wall of a fruit, enclosing the seeds.

Petals.—The leaves of the corolla.

Pistil.—The central part of a flower, composed of one or more carpels and enclosing the ovules.

Pollen.—The cells discharged from the stamens; they produce the male cells and correspond to the microspores of Spore-plants.

Pollen-sac.—A sporangium in which pollen is formed.

Pollen-tube.—The outgrowth produced from a pollen-grain on germination; it conveys the male cells to the egg-cells.

Pollination.—The bringing of pollen to the stigma or ovule.

Prothallus.—The simple sexual individual produced from the spore in Ferns and their allies.

Receptacle.—The end of the flower-stalk, bearing the various organs of the flower.

Sepals.—The leaves of the calyx.

Sorus.—A group of sporangia in a Fern.

Spermatozoid.—An actively moving male cell.

Spermophyta.—The Seed-plants.

Sporangiophore.—An organ specially adapted to bear sporangia.

Sporangium.—A sac or case in which spores are produced.

Spore.—A reproductive body in Cryptogams, usually consisting of a single cell; produced without fertilisation.

Sporogonium.—A spore-bearing fruit, developed from the fertilised egg-cell, in Mosses and Liverworts.

Sporophyll.—A leaf bearing one or more sporangia.

Stamen.—The sporophyll which bears the pollen-sacs in Seed-plants.

Stigma.—The part of the pistil which receives the pollen.

Stipule.—An outgrowth from the base of a leaf-stalk.

Style.—The part of the pistil which connects the stigma with the ovary.

Synagium.—A group of sporangia united together.

Thallus.—A plant-body in which stem and leaf are not distinguishable.

Tracheide.—A wood-cell, adapted for conveying or storing water, but not forming part of a vessel.

Vascular.—Possessing, or consisting of, wood and bast.

Vessel.—A row of cells in the wood, fused into a tube for the conveyance of water.

Whorl.—A ring of leaves or other organs, all inserted at the same level on the stem.

BIBLIOGRAPHY

BOWER, F. O.—The Origin of a Land Flora, a Theory based upon the Facts of Alternation: The Macmillan Co., 1908, 727 pp. 8vo.

A clear and well-illustrated account of the Higher Cryptogams; valuable for facts, apart from the author's theory.

CAMPBELL, D. H.—The Structure and Development of Mosses and Ferns: The Macmillan Co., 2nd edition, 1905, 657 pp. 8vo.

Another very useful illustrated book on the Higher Cryptogams, going fully into their reproduction and embryology.

COULTER, J. M., and CHAMBERLAIN, C. J.—Morphology of Gymnosperms: University of Chicago Press, 1910, 458 pp. 8vo.

By far the best "up-to-date" treatise on Gymnosperms, recent and fossil, remarkable for its excellent illustrations.

DARWIN, CHARLES.—The Origin of Species: D. Appleton & Co. 2 vols. in one, 1910, 703 pp. 8vo.

THE DARWIN-WALLACE CELEBRATION held on 1st July, 1908, by the Linnean Society of London: Linnean Society, 1908. 139 pp., 8 portraits, 8vo.

Contains, besides various addresses by Wallace, Hooker and others, reprints of the original papers by Darwin and Wallace, which preceded the publication of the Origin of Species.

DARWIN AND MODERN SCIENCE.—Edited by A. C. Seward: Cambridge University Press, 1909, 595 pp. 8vo.

The Darwin Memorial volume, consisting of a collection of 29 essays by various authors, dealing with Darwinism and Evolution from every point of view.

HOFMEISTER, W.—On the Germination, Development and Fructification of the Higher Cryptogamia, and on the Fructification of the Coniferæ. Translated by F. Currey: Ray Society, 1862, 506 pp., 65 plates, 8vo.

A classical work, laying the foundations of our scientific knowledge of these classes of Plants. All the essentials still hold good.

JUDD, J. W.—The Coming of Évolution: Cambridge University Press, 1910, 171 pp., 4 portraits, 12mo.

A popular and interesting sketch of the rise of the doctrine of Evolution, especially in relation to Geology.

KERNER, A., and OLIVER, F. W.—The Natural History of Plants: Henry Holt & Co., 1895, 2 vols., 1760 pp., 2000 figures, large 8vo.

A storehouse of all kinds of fascinating knowledge relating to every form of plant-life.

LAURENT, L.—Les Progrès de la Paléobotanique Angiospermique: Progressus Rei Botanicæ. Vol. I, 1907, 49 pp. 8vo.

An article giving the recent advances in our knowledge of fossil Angiosperms.

SCOTT, D. H.—An Introduction to Structural Botany: Part I, Flowering Plants: 7th edition, 1909, 290 pp. small 8vo.

An elementary, illustrated account of the structure of Seed-plants, on the type system.

Do. Part II; Flowerless Plants: 5th edition, 1907, 316 pp. small 8vo, Adam and Charles Black.

A similar account of the various groups of Cryptogams.

SCOTT, D. H.—Studies in Fossil Botany: 2nd edition, 1908–9, Adam and Charles Black, 676 pp. 8vo.

An illustrated account of the most important types of Palæozoic and Mesozoic plants, from an evolutionary point of view.

SCOTT, D. H.—The Present Position of Palæozoic Botany: Progressus Rei Botanicæ. Vol. I, 1907, 79 pp. 8vo.

An article giving the main results of the investigation of Palæozoic plants.

SEWARD, A. C.—Fossil Plants, for Students of Botany and Geology: Cambridge University Press. Vol. I, 1898, 452 pp.; Vol. II, 1910, 624 pp. 8vo.

A full and comprehensive work on Fossil Botany, written from a geological as well as a botanical point of view, and abundantly illustrated. In course of publication; the two volumes already published deal with all classes of Cryptogams.

STOPES, MARIE C.—Ancient Plants, being a simple account of the past Vegetation of the Earth: Blackie, 1910, 198 pp. 8vo.

A popular illustrated sketch of all branches of Fossil Botany.

WALLACE, A. R.—The World of Life: Chapman and Hall, 1910, 408 pp. 110 figures, 8vo.

Interesting, as giving the latest views of one of the founders of the theory of Natural Selection.

WIELAND, G. R.—American Fossil Cycads: Carnegie Institution, 1906, 296 pp., 50 plates and many figures, 4to.

A magnificent work, giving the full results of the author's classical researches, with comparative observations on modern Cycads, etc.

ZEILLER, R.—Les Progrès de la Paléobotanique à l'Ère des Gymnospermes: Progressus Rei Botanicæ. Vol. II, 1907, 56 pp. 8vo.

An article giving all the recent results of the investigation of the Mesozoid floras.

INDEX

*[The * refers to inscription below the illustration.]*